Metal Matrix Composites by Friction Stir Processing

Metal Matrix Composites by Friction Stir Processing

Ranjit Bauri and Devinder Yadav
Department of Metallurgical and Materials Engineering
Indian Institute of Technology Madras
Chennai, India

Butterworth-Heinemann is an imprint of Elsevier
The Boulevard, Langford Lane, Kidlington, Oxford OX5 1GB, United Kingdom
50 Hampshire Street, 5th Floor, Cambridge, MA 02139, United States

Notices
Knowledge and best practice in this field are constantly changing. As new research and experience broaden our understanding, changes in research methods, professional practices, or medical treatment may become necessary.

Practitioners and researchers must always rely on their own experience and knowledge in evaluating and using any information, methods, compounds, or experiments described herein. In using such information or methods they should be mindful of their own safety and the safety of others, including parties for whom they have a professional responsibility.

To the fullest extent of the law, neither the Publisher nor the authors, contributors, or editors, assume any liability for any injury and/or damage to persons or property as a matter of products liability, negligence or otherwise, or from any use or operation of any methods, products, instructions, or ideas contained in the material herein.

British Library Cataloguing-in-Publication Data
A catalogue record for this book is available from the British Library

Library of Congress Cataloging-in-Publication Data
A catalog record for this book is available from the Library of Congress

ISBN: 978-0-12-813729-1

Publisher: Matthew Deans
Acquisition Editor: Christina Gifford
Editorial Project Manager: Charlotte Kent
Production Project Manager: Sruthi Satheesh
Cover Designer: MPS

Typeset by MPS Limited, Chennai, India

CONTENTS

LIST OF FIGURES

LIST OF TABLES

PREFACE

This is the eighth volume in the recently launched short book series on friction stir welding (FSW) and friction stir processing (FSP). As highlighted in the preface of the first book, the intention of this book series is to serve engineers and researchers engaged in advanced and innovative manufacturing techniques. FSW was invented more than 20 years back as a solid-state joining technique. In this period, FSW has found a wide range of applications in joining of aluminum alloys. FSP was invented based on the principle of FSW. In little more than one decade, FSP has emerged as a popular and effective tool for grain refinement and microstructure modification. FSP, however, can have a far greater reach if the full potential of the process is realized through new material development by the process.

This book is focused on processing of metal matrix composites (MMCs) by FSP to highlight the potential of FSP as a tool to manufacture new materials. Apart from grain refinement, FSP has a wider capability due to the material flow and material mixing that happens during the process. This volume aims to highlight such aspects by demonstrating the ability of the process to incorporate a second phase and make MMCs. The book will cover the subject in detail and provide a good summary of the work done so far with regard to processing MMCs by FSP. It will also present a novel approach of making ductile MMCs by FSP using metal particle reinforcements. As a whole, the book aims to present a new facet of FSP to the reader community. As stated in the previous volumes, this short book series on FSW and FSP will include books that advance both the science and technology.

PREFACE

ACKNOWLEDGMENTS

All of the friction stir processing experiments of the authors reported in this book, especially Chapter 4, Processing Nonequilibrium Composite (NMMC) by FSP, were performed at the Materials Joining Laboratory, Department of Metallurgical and Materials Engineering, IIT Madras. The authors would like to express their sincere gratitude to Head of the Department and Head, Materials Joining Laboratory for providing the FSP facility and all other characterization facilities used for the work. Some of the FSP facilities and part of the work were financially supported by Naval Research Board (NRB), Govt. of India, and the authors would like to acknowledge and thank NRB for the support. The SEM and EBSD studies were carried out in the electron microscope facility created with support from Department of Science and Technology (DST), Govt. of India, under the scheme Fund for Improvement of Science & Technology Infrastructure (FIST). IIT Madras and Ministry of Human Resource & Development (MHRD), Govt. of India, are also gratefully acknowledged for providing the research scholarships and all other student infrastructure.

The authors would like to sincerely thank all colleagues who contributed to this book in some form or other. The contributions of B. Balaji and C.N. Shyam Kumar are particularly acknowledged in that regard. The authors also like to thank Dr. G.D. Janaki Ram, Associate Professor, Department of Metallurgical and Materials Engineering, IIT Madras, for his involvement and contributions to part of the work reported in Chapter 4, Processing Nonequilibrium Composite (NMMC) by FSP. Thanks are due to G. Suhas and Sai Karthik for their contributions as well. The authors would like to thank K. Rangan for his help in conducting FSP experiments. Thanks are also due to colleagues who provided timely advice and involved themselves in technical discussions. The authors would particularly like to thank Prof. Rajiv S. Mishra, Professor, Materials Science and Engineering, University of North Texas, for his encouragement to write this book and for the guidance and advice. Finally, the authors would like to thank their respective families for standing beside not only during the course of this book writing but in all other time as well.

ACKNOWLEDGMENTS

CHAPTER 1

Introduction to Metal Matrix Composites

ABSTRACT

Materials with high specific strength and stiffness have seen an increasing demand for various applications that thrive on light weighting, better fuel efficiency, and higher pay load. This has necessitated the development of a class of materials known as composites. Composite materials have received wide acceptance owing to their greater potential for use as components/products with high specific strength, stiffness, enhanced wear resistance, and better high temperature properties compared to monolithic metals and alloys. Further, they can be tailor-made to satisfy specific design requirements in a variety of applications.

Composite can be broadly defined as a material system, which comprises of a discrete phase (called reinforcement) dispersed in a continuous phase (called matrix), and which derives combination of properties, not attainable by the constituents individually, from the properties, geometry and architecture of the constituents, and from the properties of the interfaces between them.

1.1 CLASSIFICATION OF COMPOSITES

Composite materials are usually classified based on the physical or chemical nature of the matrix phase and these are as follows [1].

- Metal Matrix Composites (MMCs)
- Ceramic Matrix Composites (CMCs)
- Polymer Matrix Composites (PMCs)

There are also some reports on the advent of intermetallic and carbon matrix composites.

1.2 METAL MATRIX COMPOSITES

In a metal matrix composite (MMC) a harder and stronger phase such as a ceramic is added to a ductile metallic matrix. MMCs have gained widespread attention because of a range of attractive properties such as high specific strength and modulus, low coefficient of thermal expansion (CTE), better wear resistance, and improved high

Metal Matrix Composites by Friction Stir Processing. DOI: http://dx.doi.org/10.1016/B978-0-12-813729-1.00001-2

temperature properties. MMCs offer the advantage of combining the properties of the metallic matrix (ductility and toughness) with those of the ceramic reinforcement (high strength and stiffness). MMCs based on several matrices have been studied extensively [2–6]. The lightweight metals such as aluminum, magnesium, and titanium have emerged as the most popular choices as the matrix materials. The properties of MMCs based on these matrices can even surpass those of heavier ferrous materials. Aluminum is the most attractive matrix material particularly for aerospace and automotive industries where light weight is an important criterion. Aluminum matrix composites (AMCs) have gained immense popularity and acceptance in past three decades owing to their high strength to weight ratio and superior wear resistance. Use of low density AMCs can result in significant fuel saving and higher fuel efficiency in aerospace and automotive industries. The availability of newer and inexpensive reinforcements has fueled the growing interest in MMCs. Commonly used reinforcements in MMCs are Al_2O_3, SiC, B_4C, AlN, and Gr (graphite). Nanostructures such as carbon nanotubes (CNTs) have also been used as reinforcement in recent time. Based on the reinforcement geometry MMCs can be classified into three categories:

- Particle reinforced Metal Matrix Composites (PMMCs)
- Short Fiber reinforced Metal Matrix Composites (SFMMCs)
- Continuous Fiber reinforced Metal Matrix Composites (CFMMCs).

A schematic of the type of composites basis the reinforcement geometry is shown in Fig. 1.1.

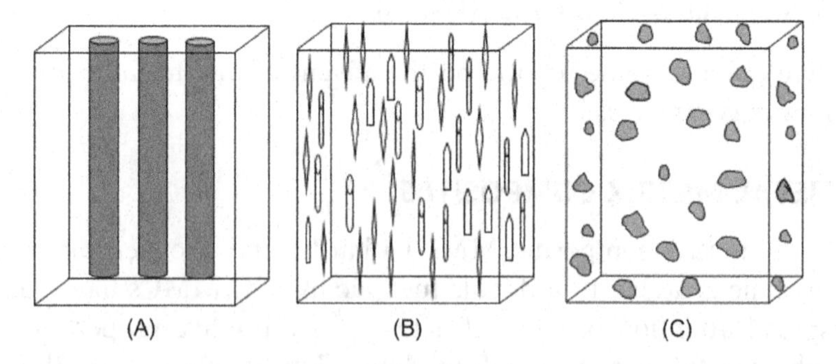

Figure 1.1 Schematic of type of composites based on reinforcement geometry, (A) fiber reinforced, (B) whisker/short fiber reinforced, and (C) particle reinforced composite.

1.2.1 Particle Reinforced Metal Matrix Composites (PMMCs)

PMMCs contain ceramic particles with an aspect ratio less than five. The shape of the reinforcing particles may vary as spherical, platelet, or any other regular or irregular shape. The most commonly used particulates are SiC and Al_2O_3 [7–9]. However, other reinforcements such as B_4C, ZrO_2, TiC, TiN, AlN, graphite, clay, and mica have also been used [2,10–16]. The size and volume fraction of particles depend upon the targeted application. Particles size of 1–20 μm and content of 10–30 vol.% is typical of many PMMCs targeted for structural applications. In certain applications such as electronic packaging particle volume fraction can be as high as 70%. The prime factors behind the widespread acceptability of PMMCs are (a) availability of a range of reinforcement at reasonable costs, (b) established manufacturing processes that can yield reproducible microstructures and properties, and (c) availability of metal working processes to shape these materials [17]. PMMCs are most commonly manufactured either by liquid metallurgy route or by powder blending and consolidation techniques. Other processing routes include reactive processing and spray codeposition.

1.2.2 Short Fiber Reinforced Metal Matrix Composites (SFMMCs)

The development of short alumina fiber reinforced aluminum MMC pistons for diesel engine saw the emergence of SFMMCs in mid 80s. This kind of fiber reinforcements are typically characterized by a high aspect ratio that is greater than five. Melt infiltration is a commonly used process for making SFMMCs. Secondary processing, such as extrusion and forging can be also applied on these composites. However, their formability is generally inferior compared to PMMCs.

Better mechanical properties are obtained when the short fibers are substituted by whiskers. These whiskers typically have diameter ≤ 1 μm with aspect ratios up to several hundreds. Though superior properties have been reported for whisker-reinforced MMCs, work in this area diminished in the early 90s with no or little commercial application. This is largely due to the health hazards associated with whiskers. Whisker fragments in the submicron size range can easily become airborne and can damage the lungs if inhaled.

1.2.3 Continuous Fiber Reinforced Metal Matrix Composites (CFMMCs)

A number of continuous fiber MMCs have been studied and some have also reported to have found certain applications [18]. The reinforcement may be multifilaments or monofilaments. *Multifilaments* are smaller diameter (~5–30 μm) fibers that can be handled as tows or bundles. Commonly used multifilaments include carbon, SiC, and several oxides. Other multifilaments (e.g., polymeric and organic fiber) are of limited interest for MMCs since most of these cannot survive the temperatures encountered during MMC fabrication. Multifilament MMCs can be fabricated by pressure infiltration. *Monofilaments* are larger in diameter (~100–150 μm) and not as flexible as multifilaments. Hence, these are used as single fibers and extra care is needed to prevent damage during handling. These composites exhibit directionality or anisotropy, i.e., variation in strength with fiber orientation.

Further, another variant of MMCs known as hybrid MMCs [19,20] has been developed in recent years. Hybrid MMCs, as the name suggests, contain more than one type of reinforcements. The hybridization could be a mix of particle and whisker, or fiber and particle, or a mixture of hard and soft reinforcements. Hybrid aluminum matrix composites containing a mixture of carbon fiber and alumina particles have been used in cylinder liners.

1.3 PROCESSING OF MMCs

Various processing routes have been in use for the fabrication of MMCs. Selection of a proper processing technique for the fabrication of MMCs is very important since the properties and cost are determined by the used fabrication method. The processing routes can be classified based on the state of the matrix, i.e., whether it is in liquid or solid or vapor state during processing. In general, methods of fabrication of MMCs can be classified under the following categories.

1.3.1 Liquid State Processing

1.3.1.1 Stir Casting

This process involves vigorous stirring of the molten metal by a mechanical impeller that creates a vortex in the molten pool. The ceramic particles in loose form are incorporated through the periphery of the vortex and the mixture is then allowed to solidify [7,21,22]. The method was originally developed by Surappa and Rohatgi for aluminum matrix composites [22]. The mechanical force, provided by the stirring action, is required to overcome the poor wettability of the reinforcement-metal system and to combine these constituents into a uniform mixture. The stirring action also helps keeping the dispersed particles in suspension. A wide range of ceramic particles can be added up to 30 vol.% in different aluminum alloys. Stir casting is the most economical processing route [23–25] of manufacturing MMCs on a commercial scale. However, a number of process parameters have to be controlled to obtain a quality product. Nonoptimum parameters lead to poor wetting and nonuniform distribution (clustering) of particles and this adversely affects the final mechanical properties of the composites. Gas entrapment by the particles leading to formation of pores is another issue that needs to be taken care by proper degassing. Stir casting technique also involves long contact between the liquid and the ceramic particles that can cause interfacial reaction. Formation of interfacial reaction products also degrades the properties of the composite. The stir casting process has been scaled up in capacity over 11,000 tons per year for commercial applications of aluminum MMCs (e.g., Duralcan, USA, and Hydro Aluminum, Norway) [25,26].

Another variant of stir casting process is known as compocasting [27,28]. In the compocasting process, the ceramic particles are added into the matrix alloy when it is in semisolid state.

1.3.1.2 Melt Infiltration Process

In the melt infiltration processes the reinforcement is first assembled to make a porous *preform*. The preform is then infiltrated with molten metal which flows through interstices and fill the pores to form the composite [18]. Infiltration of molten metal can be done with or without the application of pressure. When the liquid metal is injected by hydraulic pressure, it is known as *squeeze infiltration* [29]. The pressure can be also applied by a gas medium [30]. In the *pressureless* process, the preform is placed over an alloy ingot and the assembly is heated above the liquidus temperature of the alloy. The liquid is gravity fed through the capillary forces that are developed once the preform is wetted by the liquid. Therefore, composition of the alloy is important here for proper wetting of the preform. Major advantages of infiltration process include production of near-net shape parts with selective reinforcements. Dense composites with uniform microstructures can be produced. A limitation of this process is that the reinforcement has to be assembled as a self-supporting structure. Damages to the preform or fibers during infiltration or clustering of the reinforcement may result in heterogeneity in the final product leading to deterioration of mechanical properties of the composite. *Lanxide process* [20,31] and *PRIMEX* [32] process are examples of *pressureless infiltration* process.

1.3.1.3 Spray Deposition

In the spray deposition process the reinforcement particles are introduced into a molten stream of the matrix alloy. There are two categories of spray deposition: (i) Osprey process and (ii) thermal spray process. In the Osprey process [33] the matrix alloy is melted and subsequently atomized by inert gas jets. The reinforcement particles are then introduced into this molten stream. The mixture droplets are allowed to deposit on a substrate in the form of a metal matrix composite. A variety of shapes such as rounds, strips, or clad products can be produced provided the process parameters are properly controlled. Continuous fiber reinforced MMCs can also be fabricated by spray process. In that case, the molten metal is sprayed onto the fibers which are wrapped around a mandrel that controls the interfiber spacing. In the thermal spray process, the droplet stream is produced by continuously feeding cold metal into a zone of rapid heating.

1.3.1.4 In situ Process

In the in situ process the reinforcement and the interfaces are generated during processing itself. The *XD* process [20] is an example of in situ technique for processing MMCs. This process is essentially a solid-state self-propagating reaction synthesis in which the constituent powders are mixed and ignited to generate the reinforcement through a chemical reaction. For example, $Al + TiB_2$ composites can be processed by blending and heating of powders of Al, Ti, and B. The exothermic nature of the reaction makes it self-propagating. It is possible to obtain composites having submicroscopic dispersion of reinforcing particles by this process. An alternative and economical method of producing $Al + TiB_2$ in situ composite is to use B and Ti containing salts such as KBF_4 and K_2TiF_6, respectively, as precursors to form the TiB_2 reinforcement in the Al melt [34,35]. The overall reaction can be written as follows.

$$Al + KBF_4 + K_2TiF_6 \rightarrow KAlF_4 + K_3AlF_6 + TiB_2 \quad (1.1)$$

The spent salt is decanted before pouring and the melt solidifies in the form of an Al-TiB_2 composite.

In some processes, the liquid metal is introduced into an oxidizing atmosphere and progressively oxidized. An example of such a process is the direct melt oxidation process known as "*DIMOX*" [20]. In situ composites have several advantages such as a clean interface as the reinforcing phase is thermodynamically stable, fine particle size, and homogeneous distribution of particles. Moreover, the spacing or size of the reinforcement could be manipulated. However, controlling the kinetics or the shape of the reinforcement is difficult. The choice of materials system is also limited.

1.3.2 Solid-State Processing

1.3.2.1 Powder Metallurgy

Powder metallurgy is the primary solid-state synthesis technique used in the fabrication of both particle and whisker-reinforced MMCs. It involves blending of prealloyed or elemental powder of the matrix with ceramic whiskers or particulates and consolidation by die compaction, canning, powder forging or extrusion [36]. In some cases, hot pressing or hot isostatic pressing (HIP) is also employed. Several commercial manufacturers such as ALCOA and Advanced Composite Materials Inc. have used powder consolidation technology to fabricate variety of composites. Ceramic fibers reinforced composites have also been processed, however, it is difficult to achieve uniform fiber spacing. MMCs processed by powder metallurgy are generally extruded to minimize the defects like porosity and achieve homogeneous particle distribution. Extrusion, however, can result in alignment of particles along the extrusion axis leading to formation of particle rich bands parallel to the extrusion axis.

1.3.2.2 High-Energy (High Rate) Process

In this approach, high energy (mechanical or electrical energy) is applied rapidly to consolidate a metal–ceramic powder mixture [7,37]. The application of high energy at high rate results in rapid heating of the powder mix inside the die. High temperature and short processing time offer the advantage of controlling phase transformations and grain growth. Al–SiC composites have been successfully processed by this technique.

1.3.2.3 Diffusion Bonding

Fiber (monofilament) reinforced MMCs can be produced by diffusion between two layers of matrix that hold the aligned fibers in between them. The technique is known as foil-fiber-foil method. Diffusion bonding can also be achieved by evaporating and depositing thick layers of matrix material onto the fibers [38]. 6061Al-Boron fiber composites have been processed by foil-fiber-foil route. However, majority of the work in this area is focused on titanium based composites. Diffusion bonding titanium is not difficult as the surface oxide layer dissolves at higher temperatures. Long fibers reinforced Ti composites have been produced commercially by placing arrays of fibers between thin foils followed by hot pressing. One of the major problems in this process is the formation of reaction products at fiber/metal interface. The foil-fiber-foil route is generally cumbersome and it is difficult to obtain high fiber volume fractions and homogeneous fiber distributions. It is also difficult to produce complex shaped products using this process.

1.3.3 Vapor State Processing

Physical vapor deposition (PVD) is one of the prominent techniques used in the fabrication of MMCs [20]. In this process, fibers are continuously passed through a chamber filled with vapors of the matrix material. Relatively thick coatings are produced when condensation takes place over the fiber. The vapor can be generated by high intensity electron beam (EB-PVD) directed on to the end of a solid feedstock. The composite is finally fabricated by assembling the coated fibers into bundle or arrays and consolidating them by hot pressing or hot isostatic pressing. One of the biggest advantages of this technique is that a wide range of matrix composition can be used.

Some new processing routes have also been developed in recent time. Friction stir processing (FSP) is a solid-state thermomechanical process that has emerged as an effective tool for grain refinement and microstructure modification. The material flow and mixing action during FSP coupled with the frictional heat enable incorporation of secondary particles in the processed zone and thus allows fabrication of metal matrix composites. Fabrication of MMCs by FSP is the subject matter of this book and will be described in details in the subsequent chapters.

1.4 PROPERTIES OF MMCs

The physical and mechanical properties of MMCs are generally found to be superior compared to those of the unreinforced alloys. One of the major advantages of MMCs is the high specific properties. The specific strength and modulus of MMCs are shown in Fig. 1.2 in

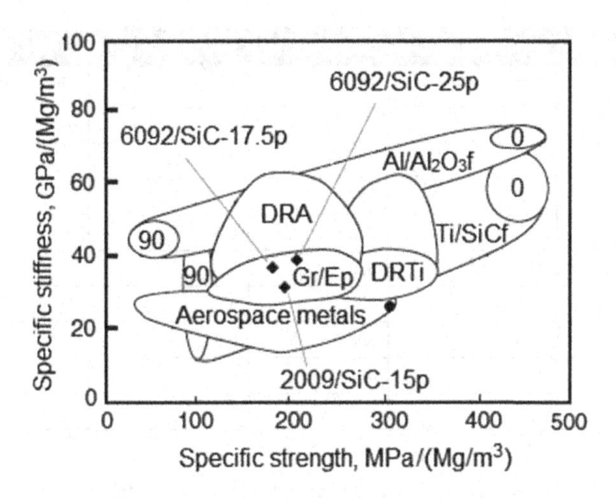

Figure 1.2 Specific strength versus modulus of MMCs in longitudinal (0°) and transverse (90°) direction in comparison to conventionally used aerospace metals.

comparison with conventional aerospace metals. Although fiber reinforced composites show much better specific strength and stiffness in longitudinal direction (0 degree), these are generally found to be inferior in the transverse direction (90 degrees) as shown in Fig. 1.2. Discontinuously reinforced Al (DRA) composites, on the other hand, exhibit isotropic properties and also display better specific properties compared to light-weight monolithic metals and alloys such as Al, Mg, or Ti, and steels. Here we will briefly review the mechanical properties of Al based metal matrix composites.

In MMCs the ceramic reinforcement is added with the objective of increasing the strength and stiffness. This can be done in two ways; one, by incorporating reinforcements, which possess high intrinsic strength and stiffness, or alternatively by addition of reinforcements, which can modify the microstructure in such a way so as to increase the strength and stiffness. These two aspects can also be combined together to achieve the same objective.

The Young's modulus of metals can be improved significantly by ceramic particles reinforcement [39]. The improvement is particularly attractive for low-stiffness metals such as aluminum and magnesium. The addition of ceramic reinforcement generally improves the yield strength. However, reaction between matrix and ceramic that forms unwanted reaction product at the particle–matrix interface can degrade the strength particularly at higher reinforcement content. The

Table 1.1 Tensile Properties of Different Al Alloys and Their Composites [39]

Material	0.2% PS (MPa)	UTS (MPa)	Elongation (%)	Elastic Modulus (GPa)
6061Al	276	310	20	69
6061Al-10%Al_2O_3	296	338	7.5	81
6061Al-15% Al_2O_3	317	359	5.4	87
8090Al	335	450	9.0	80
8090Al-13%SiC	455	520	4	101
A356Al	200	276	6	75
A356-10%SiC	287	308	0.6	82
2014Al	414	483	13	73
2014Al-20% Al_2O_3	483	503	1.0	101

trend in mechanical properties especially, the specific strength and specific stiffness, puts MMCs at a superior level compared to monolithic alloys. Table 1.1 shows the typical tensile properties of some Al alloys and their composites.

Although the improved modulus and strength of MMCs are well documented, the physical basis for the improvement in these properties is still open to debate. There are two main approaches viz. continuum approach and micromechanistic approach, for interpreting the mechanical properties of MMCs. In the micromechanistic approach, the understanding is based on the knowledge of the deformation processes at the atomic level. In the continuum approach, it is assumed that the material property can be described by global parameters. The continuum model has been successfully applied to predict the elastic modulus, CTE, and other physical properties of particulate and continuous fiber reinforced composites. These models predict an increase in strength with increase in volume fraction and particle aspect ratio. However, the effect of particle size is not taken into account. Several mechanisms have been postulated to account for the strengthening observed in particle reinforced MMCs [40]. The main strengthening mechanisms in MMCs are briefly described in the next section.

1.5 STRENGTHENING MECHANISMS IN MMCs

1.5.1 Dislocation Strengthening

The large difference in thermal expansion coefficient between aluminum and ceramic particles results in generation of dislocations while cooling from the processing temperature. The CTE of most Al alloy is typically $\sim 21 \times 10^{-6}$°C^{-1} and that of ceramic like SiC is $\sim 4.5 \times 10^{-6}$°C^{-1} [41]. Therefore, even a small change in temperature will generate thermal residual stresses near the reinforcement, which may be partially relaxed by generation of dislocations. The enhanced dislocation density in the matrix of the composite leads to an increase in yield strength. The dislocations density (ρ) after quenching is dependent on particle size (d_p) and volume fraction (F_v), thermal mismatch ($\Delta\beta$) and the temperature change (ΔT). The contribution of dislocations to strength (σ_d) can be estimated by the following expression.

$$\sigma_d = \alpha G b \rho^{1/2} \tag{1.2}$$

where G is shear modulus, b is Burgers vector, and α is a constant having value between 0.5 and 1. The dislocation density due to thermal mismatch ($\Delta\beta$) is given by:

$$\rho = \frac{12 \Delta T \Delta \beta F_V}{b d_p (1 - F_V)} \tag{1.3}$$

The above approach assumes that the dislocations are uniformly distributed and all the generated dislocations contribute to σ_d.

1.5.2 Orowan Strengthening

The glissile dislocations can bow between impenetrable particles and bypass them leaving behind loops known as Orowan loops. Orowan bypassing of particles by dislocations can enhance the strength of the composite. If the particles are assumed to be equiaxed, the Orowan strengthening (σ_o) is given in the simplest form by:

$$\sigma_O = \frac{Gb}{\lambda} \tag{1.4}$$

where λ is the interparticle distance which depends on the particle size and volume fraction.

For a cubic arrangement of spherical particles the interparticle spacing is given by Ref. [42].

$$\lambda = d_p \left[\left(\pi / 4 F_V \right)^{1/2} - 1 \right] \tag{1.5}$$

A decreasing particle size and increasing volume fraction leads to decreasing interparticle spacing and hence, increasing strength.

1.5.3 Grain Size Strengthening

The particles act as sites for nucleation during solidification or recrystallization of MMCs. This results in higher nucleation rate and thus, finer grain size. In a particulate composite having particles bigger than a certain size (>1 μm), the recrystallization is enhanced, a phenomena known as particle-stimulated nucleation (PSN). The resultant grain size (D) is given by Ref. [43]

$$D \approx d_p F_V^{-1/3} \tag{1.6}$$

As the grain size of the matrix of the composite is considerably finer than that of an unreinforced alloy, the increase in yield strength (σ_y) of the MMCs can be estimated by the Hall–Petch relationship

$$\sigma_y = \sigma_i + kd^{-1/2} \tag{1.7}$$

where σ_i is the friction stress or the overall lattice resistance to dislocation motion and k is a constant, typically 0.1 $MNm^{-3/2}$.

1.5.4 Work Hardening

During plastic deformation, the unreinforced alloys generally exhibit a gradual increase in the yield strength that is connected with the long-range dislocation glide. The presence of the reinforcement causes interruption to the material flow leading to dislocation generation. Dislocation tangles can form around the particles at low strains due to this plastic incompatibility. As the deformation continues, geometrically necessary dislocations (GNDs) are generated at higher strains [44]. The presence of the reinforcement thus enhances the work hardening of the matrix through forest hardening and other relevant dislocation mechanisms. There may also be other factors such as load transfer and Orowan loops contributing to the work hardening of composites.

It should be noted here that the above contributions to MMC's strength may be superimposed by alloying element effects, such as solution or precipitation hardening. Indeed, there appears no simple way of summing up various strength contributions, since various mechanisms will interact with each other.

1.6 APPLICATIONS OF MMCs

As pointed out before, MMCs have been manufactured on commercial scale for various industrial applications. MMCs are attractive for aerospace, automotive, and leisure industries due to their favorable properties. With the advent of inexpensive discontinuous reinforcements and availability of various processing routes, the use of discontinuously reinforced MMCs is increasing. For a comprehensive overview of MMCs in industry one can refer to [45]. Some of the major applications of MMCs are summarized below.

MMCs have found applications in various automotive parts such as drive shafts, disc brake rotors, pistons, and cylinder bores in engines. Fig. 1.3 shows brake rotors made from Al MMCs [46,47]. MMC cylinder bores used by Honda are made by infiltrating preforms of chopped Safiil and graphite fibers. Similar processing routes have been used to make selectively reinforced cylinder bores for Porsche Boxter and 911 Carrera [48]. Cast iron cylinder liners have been replaced by Al composites, e.g., 6061 Al/Al_2O_3 composites produced by stir casting and direct seamless extrusion [48]. Automotive tire studs made of discontinuously reinforced Al (DRA) composite is a significant market component of MMCs. In the aerospace industry MMCs find applications in structural, propulsion, and subsystem components. Ventral fins, fuel access door covers on F-16 aircrafts and rotor blade sleeves and swash plates on Eurocopter EC120 and N4 helicopters are some of the structural applications of MMCs in aerospace. Applications in aeropropulsion system

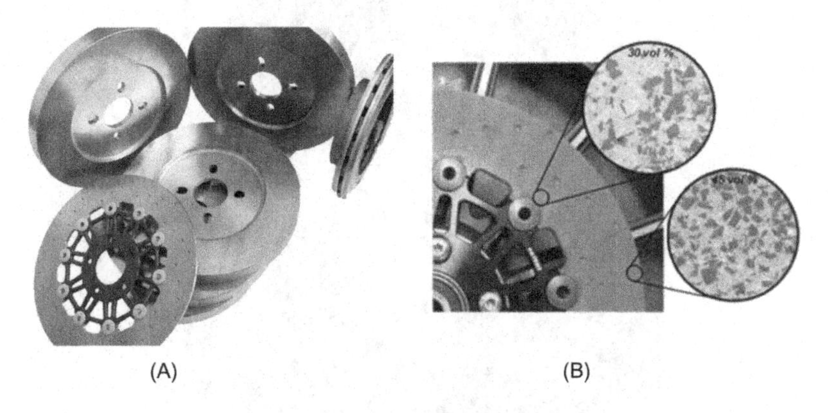

Figure 1.3 (A) Brake rotors made of Al MMC. (B) A gradient microstructure with high ceramic loading on the outer periphery where the disc moves fastest and generates maximum heat and lower loading inward to maintain flatness and minimize fade [46,47].

include fan exit guide vanes, nozzle actuator links. In the subsystem category, DRA composites find application in avionic and ammunition racks in Lockheed-Martin, hydraulic manifolds in Boeing/Bell helicopters. MMCs have found applications, though limited, in space systems as well. MMC tubes made of 6061 Al reinforced with B monofilaments have been used in space shuttle orbiter. Tows made of 6061Al/Gr fiber composite have been used in Hubble Space Telescope antenna waveguide mast. DRA composites are also used in thermal management systems of space shuttles, e.g., DRA panel is used as heat sink between two printed circuit boards (PCB).

MMCs are used in recreational and infrastructure applications as well. Recreational goods such as tennis racket, bicycle frames, skies, fishing rods, etc. have been made of MMCs. Al/B_4C composites for nuclear waste containers, Al/Saffil composites for overhead power transmission lines are examples of some of the infrastructure applications of MMCs. The transmission lines, 3M Aluminum Conductor Composite Reinforced (ACCR) high-capacity transmission conductor, are made of an Al MMC core surrounded by Al–Zr outer wires (Fig. 1.4). These are light weighted and expected to provide a 200–300% increase in the power carrying capacity [48].

Figure 1.4 Cross-section of a 3M aluminum conductor composite reinforced (ACCR) high-capacity MMC conductor used in overhead transmission lines [48].

REFERENCES

[1] M. Balasubramanian, Composite Materials and Processing, CRC Press, New York, 2014.

[2] M.K. Surappa, P.K. Rohatgi, J. Mater. Sci. 16 (1981) 983–993.

[3] T.S. Srivatsan, T.S. Sundarshan, E.J. Lavernia, Prog. Mater. Sci. 39 (1995) 317–409.

[4] R.J. Arsenault, L. Wang, C.R. Feng, Acta Metall. Mater. 39 (1991) 47–57.

[5] M. Gupta, M.K. Surappa, Key Eng. Mater. 104–107 (1995) 259–274.

[6] M. Gupta, F. Mohamed, E.J. Lavernia, T.S. Srivatsan, J. Mater. Sci. 28 (1993) 2245–2259.

[7] P.K. Rohatgi, R. Asthana, S. Das, Int. Met. Rev. 31 (1986) 115–139.

[8] P.K. Rohatgi, S. Ray, R. Asthana, C.S. Narendranath, Mater. Sci. Eng. A 162 (1993) 163–174.

[9] S. Mohan, V. Agarwala, S. Ray, Wear 140 (1990) 83–92.

[10] D.C. Halverson, A.J. Pyzik, I.A. Aksay, W.E. Snowden, J. Am. Ceram. Soc. 72 (1989) 775–780.

[11] H. Nayeb-Hahemi, D. Shan, Mater. Sci. Eng. A 266 (1999) 8–17.

[12] A. Sato, R. Mehrabian, Metall. Trans. B 13 (1976) 443–451.

[13] B.P. Krishnan, P.K. Rohatgi, Met. Technol. 11 (1984) 41–44.

[14] T.P. Murali, M.K. Surappa, P.K. Rohatgi, Metall. Trans. B 13 (1982) 485–494.

[15] F. Rana, D.M. Stefanescu, Metall. Trans. A 20 (1989) 1564–1566.

[16] A. Banerji, P.K. Rohatgi, J. Mater. Sci. 17 (1982) 335–342.

[17] P. Divecha, S.G. Fishman, S.D. Karmarkar, JOM 33 (1981) 12–17.

[18] S. Suresh, A. Mortensen, A. Needleman, Fundamentals of Metal Matrix Composites, Butterworth-Heinemann, MA, 1993, p. 297.

[19] A.K. Dhingra, JOM 38 (1986) 17.

[20] T.W. Clyne, F.R. Jones, Metal matrix composites: matrices and processing, in: A. Mortensen (Ed.), Concise Encyclopedia of Composite Materials, second ed., Elsevier Publications, 2006.

[21] S. Ray, Bull. Mater. Sci. 18 (1995) 693–709.

[22] M.K. Surappa, P.K. Rohatgi, Met. Technol. 5 (1978) 358–361.

[23] W.H. Hunt Jr., Mater. Sci. Forum 331–337 (2000) 71–84.

[24] J.U. Ejiofor, R.G. Reddy, JOM 49 (1997) 31–37.

[25] M.D. Skibo, D.M. Schuster, Materials Technol. 10 (1995) 243–247.

[26] B. Maruyama, JOM 51 (1999) 59–61.

[27] P.K. Rohatgi, N. Gupta, A. Daoud, ASM Handbook, vol. 15, Casting, 2008, pp. 1149–1164.

[28] V. Laurent, P. Jerry, G. Regazzoni, D. Apelian, J. Mater. Sci. 27 (1992) 4447–4459.

[29] A. Mortensen, M.N. Gungor, J.A. Cornie, M.C. Flemmings, JOM 38 (1986) 30–35.

[30] A. Mortensen, V.J. Michaud, M.C. Flemmings, JOM 45 (1993) 36–43.

[31] A.W. Urquhart, Mater. Sci. Eng. A 144 (1991) 75–82.

[32] J.J. Stephens, J.P. Lucas, F.M. Hosking, Scripta Metall. 22 (1988) 1307–1312.

[33] J. White, T.C. Wills, Mater. Des. 10 (1989) 121–127.

[34] L. Lu, M.O. Lai, F.L. Chen, Acta Mater. 45 (1997) 4297–4309.

[35] S. Kumar, M. Chakraborty, V.S. Sarma, B.S. Murty, Wear 265 (2008) 134–142.

[36] V.V. Bhanu Prasad, K.S. Prasad, A.K. Kuruvilla, A.B. Pandey, B.V.R. Bhat, Y.R. Mahajan, J. Mater. Sci. 26 (1991) 460–466.

[37] G. Elkabir, L. Rabenberg, C. Persad, H.L. Marcus, Scripta Metall 20 (1987) 1411–1416.

[38] P.D. Nicolaou, H.R. Piehler, S.L. Semiatin, Metall. Mater. Trans. A 26 (1995) 1129–1139.

[39] D.J. Lloyd, Int. Mater. Rev. 39 (1994) 1–23.

[40] T.W. Clyne, P.J. Withers, An Introduction to Metal Matrix Composites, Cambridge University Press, 1993, p. 84.

[41] R.U. Vaidya, Z.R. Xu, X. Li, K.K. Chawla, A.K. Zurek, J. Mater. Sci. 29 (1994) 2944–2950.

[42] A.K. Redsten, E.M. Klier, A.M. Brown, D.C. Dunand, Mater. Sci. Eng. A 201 (1995) 88–102.

[43] F.J. Humphreys, M. Hatherly, Recrystallization and Related Annealing Phenomena, Pergamon Press, Oxford, 1995.

[44] M.F. Ashby, Phil. Mag. 14 (1966) 1157–1178.

[45] A. Evans, C.S. Marchi, A. Mortensen, Metal Matrix Composites in Industry: An Introduction and a Survey, Kluwer Academic Publishers, Dordrecht, Netherlands, 2003.

[46] S. Jensen, OEM Off-Highway, October 2012.

[47] http://www.relinc.net/advanced-materials/metal-matrix-composites/lightweight-brakes/.

[48] D.B. Miracle, Compos. Sci. Technol. 65 (2005) 2526–2540.

CHAPTER 2

Introduction to Friction Stir Processing (FSP)

ABSTRACT

This chapter presents an overview of the friction stir processing (FSP) technique. A brief introduction of the working principle of FSP and processing of light metals, particularly aluminum and magnesium, is presented. The microstructural aspects with emphasis on the microstructure evolution during the process have been also discussed.

Friction stir welding (FSW) which was developed at The Welding Institute in the United Kingdom [1], is considered to be the most significant development in the area of metal joining in past few decades. The process is also regarded as a "green" technique, due to its energy efficiency, environment friendliness, and versatility [2,3]. It does not involve use of any filler material or cover gas or flux during welding. The process has been commercially used to join high-strength aerospace aluminum alloys and other metallic alloys that are difficult to weld by conventional fusion welding techniques. Further, FSW overcomes the common problems associated with fusion welding such as cracking, porosity, and alteration of alloy chemistry.

2.1 FRICTION STIR PROCESSING

FSP was developed based on the principle of FSW [4]. The working principle of FSP is similar to FSW and is fairly simple. During FSP a specially designed, nonconsumable tool, which has two components namely a pin and a shoulder, is used. The tool is made to rotate at high speed and a downward force is applied to the tool so that the pin plunges into the base material and the shoulder just touches the surface. A schematic of the process is shown in Fig. 2.1.

As the tool rotates, large frictional heat is generated at the tool–workpiece interface that takes the temperature of the material to a range where it becomes plastic. The tool is then made to traverse at a desired speed. As the tool moves forward the plasticized material flows around the pin and is consolidated by the forward movement of the shoulder leaving behind a processed zone. The plasticized material

Metal Matrix Composites by Friction Stir Processing. DOI: http://dx.doi.org/10.1016/B978-0-12-813729-1.00002-4

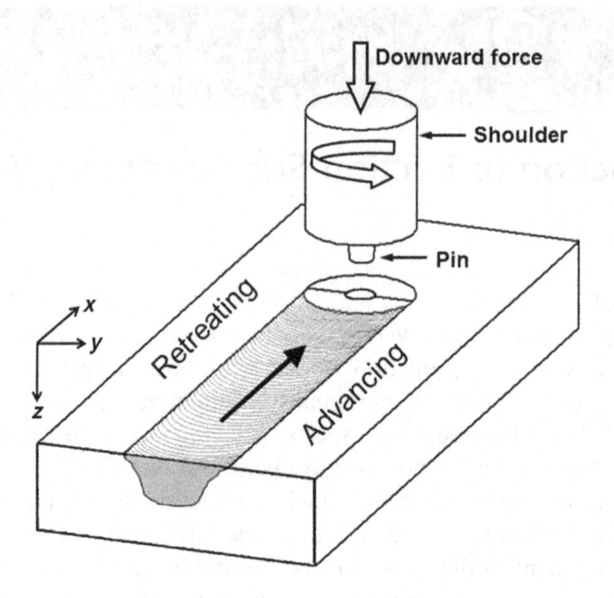

Figure 2.1 Schematic of friction stir processing.

flows in a complex way around the tool from one side to another due to the simultaneous rotational and linear motion of the tool. The tool is generally given a tilt of 1–3 degrees so that the trailing edge has a downward forging action. The side in which the tangential velocity of the tool surface is parallel to the traverse direction is defined as advancing side and the other side is defined as the retreating side as marked in Fig. 2.1. In principle, FSP is a thermomechanical process which involves severe plastic deformation of the material at elevated temperatures, typically greater than $0.5T_m$. Temperatures as high as $0.9T_m$ have been reported during FSP and since there are no evidences of melting it is considered as a solid state process [2,3].

The process was originally subjected to aluminum alloys and other soft metals such as magnesium. However, in the recent past it has been successfully applied to a variety of materials including copper, brass, iron, steel, and high entropy alloys [5–9]. Even hard to deform hexagonal close pack metals and low ductility alloys are processed by this technique since the deformation occurs at elevated temperature [10]. FSP has emerged as an effective tool for material processing, material development, and can be used for a variety of microstructural modifications. The thermomechanical nature of FSP is effectively used to refine the grain size in various metals and alloys. Starting with a coarse grained material, fine and ultrafine grained microstructures have been produced [11–12]. Even nanostructured materials have been produced by cooling the plate behind the tool to prevent the growth of dynamically recrystallized grains [13]. Table 2.1 summarizes the process parameters and the grain sizes achieved during FSP of some of the aluminum and magnesium alloys. The ultrafine grained aluminum alloys processed by FSP have also exhibited superplasticity both at high strain rates and low temperatures [14–16]. Liu and Ma reported a superplasticity of 350%–540% in the temperature range of 200–350°C with a highest strain rate of $1 \times 10^{-2}\ s^{-1}$ in FSPed Al-Zn-Mg-Cu alloy [17]. Grain boundary sliding was found to be the dominant mechanism in the superplastic deformation.

Apart from grain refinement, FSP has found its application in other areas such as modification of cast alloy microstructure, eliminating casting defects and processing a variety of composites including surface, nano and in situ ones. The material stirring action of FSP is effectively used in the cast alloys to homogenize the as-cast microstructure and eliminate the casting defects. The cast alloys suffer from various casting defects such as microporosity, dendritic microstructure, elemental segregation in the interdendritic region, and nonuniform

Table 2.1 Summary of Grain Size in the Nugget in FSP/FSW of Al and Mg Alloys

Material	Rotation speed (rpm)	Traverse speed (mm/min)	Grain size (μm)	Reference
7075 Al	–	127	2–4	[18]
6061 Al	300–1000	90–150	10	[19]
Al-Li-Cu	–	–	9	[20]
AZ91 (Mg-Al-Zn)	400	100	15	[21]
AZ31 (Mg-Al-Zn)	600	60	17	[22]
AZ61	1200	25–30	0.3	[23]
7075 Al	350, 400	102, 152	3.8, 7.5	[15]
6063 Al	360	800–2450	5.9–17.8	[24]
Al-4Mg-1Zr	350	102	1.5	[16]
2024 Al	200–300	25.4	2–3.9	[25]
Pure Al	640	150	3	[26]
2024 Al with Liq. N_2 cooling	650	60	0.5–0.8	[27]
7075 Al cooling with H_2O, dry ice and methanol	1000	120	0.1	[28]
AZ31(Mg-Al-Zn) with Liq. N_2 cooling	1200	28–33	0.1–0.3	[12]

distribution of the second phase particles which limit the mechanical properties of the cast alloys especially ductility, toughness, fatigue life, and the corrosion resistance. FSP effectively breaks the dendritic microstructure and leads to a homogeneous microstructure. Elimination of the casting defects improves both strength and ductility of the cast alloy [29–30]. The beauty of the process is that it does not change the shape and size of the base material and can be carried out selectively on a part of an engineering component for site specific property enhancement, without affecting the properties in the rest of the material.

FSP also offers the possibility of incorporating second phase particles into a material to process composites [31]. A variety of composites have been processed including surface composite, nanocomposite, intermetallic reinforced in situ composites, and hybrid composites. These will be discussed comprehensively in subsequent chapters of this book. Apart from fabricating composites, FSP is also effectively adopted to solve the age old issue of particle segregation along the grain boundaries in the in situ composites [32].

2.2 MICROSTRUCTURAL ZONES

The thermomechanical nature of FSP involves localized intense plastic deformation at elevated temperature. The combination of high temperature and strain modifies the initial microstructure in several ways and gives rise to distinct microstructural zones namely, stir zone (SZ), thermomechanically affected zone (TMAZ), and heat affected zone (HAZ). These zones are shown in Fig. 2.2 for friction stir processed (FSPed) Al. The zones exist both on the advancing and the retreating side, however, they may not have the same width on either side. The size of each zone depends on a number of factors including the tool rotation speed, traverse speed, the tool pin profile, plunge depth, base material thickness, and its thermal conductivity [2]. Each zone

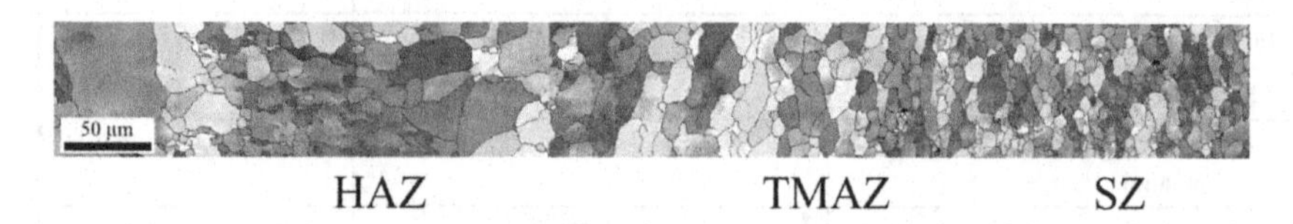

Figure 2.2 The microstructural zones in friction stir processed Al.

experiences different combination of strain rate and temperature and have different microstructural features.

2.2.1 Stir Zone

Stir zone (SZ) is the region which experiences the maximum temperature and strain. Also, the SZ undergoes maximum grain refinement and is also called as nugget zone or dynamically recrystallized zone. This zone is characterized by equiaxed fine grains. The volume of the SZ depends on the tool dimensions. Typically the width of the shoulder decides the width of the SZ and the pin length decides the depth. Because of nonuniform deformation during FSP, the advancing side has a relatively sharp interface between the SZ and TMAZ than the retreating side. The grain size is not the same throughout the volume of the SZ with variations occurring from top to bottom and also across the zone from advancing to the retreating side. Mahoney et al. reported that the average grain size ranges from 3.2 μm at the bottom to 5.3 μm at the top and from 3.5 μm at the retreating side to 5.1 μm on the advancing side in FSPed Al 7050 [33]. This is generally attributed to the temperature and strain gradients within the SZ due to different plastic flow patterns on advancing and retreating sides [34].

2.2.2 Thermomechanically Affected Zone

TMAZ exists around the volume of the SZ. The size of this zone varies from several micrometers to millimeters. TMAZ experiences lesser temperature compared to SZ and the strain rate in this zone is too low to cause dynamic recrystallization. The zone is characterized by deformed and elongated grains with high dislocation density inside them. The transition from the SZ to the TMAZ is usually not sharp and there is a zone where the microstructure changes from the recrystallized grains to the deformed grains. The width of the TMAZ zone is material dependent. Soft metals such as aluminum and magnesium have wider zones compared to harder materials such as steel and titanium.

2.2.3 Heat Affected Zone

HAZ exists between TMAZ and base material. This zone experiences no strain and is subjected to only thermal cycles which can lead to some grain growth. Due to high thermal conductivity of Al alloys grain growth occurs and this zone is characterized by grains bigger than the base metal.

In some cases it retains the grain size of the base metal. Mahoney et al. defined HAZ as a zone experiencing temperature above 250°C for heat treatable Al alloys [35]. During the tensile tests failure usually occurs in the HAZ as the mechanical properties are poor in this zone.

2.3 EFFECT OF PROCESS PARAMETERS

Rotation speed and traverse speed of the tool are the two most important process parameters which significantly affect the microstructure and hence, the mechanical properties of the processed material. The ratio of tool rotation speed to the traverse speed (ω/v ratio) decides the heat input into the material. As the rotation speed of the tool is increased the material stirring (deformation) becomes more intense and more frictional heat is generated at the tool–plate interface, and hence the temperature of the workpiece increases. At the same time as the traverse speed decreases, the tool stays in an area for longer time and consequently that volume experiences more heat. Therefore the heat input into the material is directly proportional to ratio of rotational to traverse speed and hence, the ratio decides the final grain size in the stir zone [2,26]. Higher ω/v ratio will lead to higher heat input and consequently gives rise to coarse grains while a lower ratio will lead to finer grains in the SZ. However, there exists a minimum ω/v ratio for a given material, below which the heat produced by friction and stirring is not sufficient to soften and plasticize the material and it may not flow continuously around the rotating tool thereby leaving defects. This minimum ratio will also decide the minimum achievable grain size in a material. It should be noted that too low a ratio may also lead to tool breakage as the material may not be soft enough to flow. One of the ways to achieve a finer grain size is by cooling the plate behind the tool to prevent growth of the dynamically recrystallized grains [13].

Chang et al. showed that the strain rate during FSP can be estimated by considering a torsion type deformation with the following equation [36].

$$\dot{\varepsilon} = \frac{R_m 2\pi r_e}{L_e} \tag{2.1}$$

where R_m is the average material flow rate (assumed to be half the pin rotation speed) and r_e and L_e are the effective (or average) radius and depth, respectively, of the dynamically recrystallized zone. Buffa et al. theoretically showed that the effective strain distribution was nonsymmetric about the weld line during FSW and maximum strain existed on the advancing side [37]. Maximum strain rate of 160 s^{-1} was predicted by Nandan et al. by numerical simulation confirming intense plastic deformation during FSW [38].

2.4 MATERIAL FLOW DURING FSP

The material flow during FSP is complex and the exact understanding of its role on the deformation process is rather limited. There are many factors that can influence the material flow during FSP. These factors include tool geometry (pin and shoulder design and relative dimensions of pin and shoulder), processing parameters (tool rotation speed and direction, i.e., clockwise or counter-clockwise, traverse speed, plunge depth, and tool tilt angle), material types, workpiece temperature, etc. It is very likely that the material flow within the volume of the SZ during FSP consists of several independent deformation processes [2]. The material flow behavior in FSW was investigated using a marker insert technique in 2195Al-T8 alloy [39]. In this method markers made of 5454Al-H32 were inserted in the path of the rotating tool. The final position of the inserts after welding was revealed by milling off successive layers (0.25 mm) of material from the top surface of the weld followed by etching with Keller's reagent and metallographic inspection. It was observed that the flow was not symmetric about the weld centerline. The material flows from the retreating side to the advancing side and as the tool moves ahead in the advancing side, the deformed material is deposited in the retreating side. Also, the material is pushed downward on the advancing side and moves upward in the retreating side. The authors also reported that the amount of vertical displacement of the bottom marker in the retreating side was inversely proportional to the weld pitch (the tool advance per rotation).

2.5 TEMPERATURE DISTRIBUTION

The temperature of the material rises during FSP largely due to the frictional heat generated at the tool–plate interface. The adiabatic heating due to the plastic deformation also causes some temperature rise in the material. The material experiences peak temperatures on top of the SZ near the tool shoulder. Temperature has a major role in affecting the final microstructure in the SZ. The final grain size in various zones, amount of recrystallization, dislocation recovery, precipitate dissolution, and coarsening are all affected by the temperature distribution across the processed zone [2,11]. These aspects in turn affect the mechanical properties of the material. However, understanding in this regard is limited and there is no literature on experimental measurement of exact temperature distribution within the SZ during FSP. Due to the intense material flow it is difficult to place thermocouple near the tool and temperature measurement within the SZ is thus difficult. Therefore, the temperature of the stir zone during FSW has been recorded by placing thermocouple in the regions adjacent to the rotating pin [35,40–41]. Alternatively, the SZ temperature is estimated from the microstructure of the weld [17]. Based on their observation of dissolution of larger precipitates and reprecipitation in the weld center, Rhodes et al. concluded that the maximum process temperatures are between 400°C and 480°C in FSW of 7075Al-T651 [18]. However, other studies showed that some of the precipitates were not dissolved during welding and suggested that the temperature rises to roughly about 400°C in FSW of 6061 Al [42]. The peak temperature in the weld zone increases with increasing ω/v ratio. A peak temperature greater than 550°C was observed in FSW of 5083Al at a high ω/v ratio [43]. Chao et al. reported that about 95% of the heat generated from friction is transferred to the workpiece and only 5% flows into the tool [44].

The maximum temperature of the stir zone is a strong function of the rotation speed (ω, rpm) of the tool and the rate of heating depends on the traverse speed (v, mm/min) of the tool. A general relationship between maximum welding temperature and FSW parameters (ω and v) for various aluminum alloys was given by Arbegast and Hartley [45].

$$\frac{T}{T_{\mathrm{m}}} = C\left(\frac{\omega^2}{v \times 10^4}\right)^{\eta} \tag{2.2}$$

where the exponent η was reported to be in the range of 0.04–0.06, the constant C is between 0.65 and 0.75, and T_{m} (°C) is the melting point of the material. The authors showed that the advancing side experienced slightly higher temperature than the retreating side. Based on various studies on aluminum it was concluded that the maximum temperature within the SZ during FSW/FSP is below the melting point of aluminum and it can go up to $0.9T_{\mathrm{m}}$ depending on the ω/v ratio.

2.6 MICROSTRUCTURE EVOLUTION DURING FSP

FSP is a thermomechanical process and gives rise to fine and equiaxed grains in the stir zone. Various processes such as recovery, recrystallization, precipitation, dissolution, and local softening may occur during the process. It is generally believed that the grain refinement happens due to the dynamic recrystallization (DRX) process. Various forms of dynamic recrystallization have been proposed as the mechanism of grain refinement during FSP of various Al alloys [2]. These include continuous dynamic recrystallization (CDRX), discontinuous dynamic recrystallization (DDRX), geometric dynamic recrystallization (GDRX), and dynamic recovery (DRV).

Due to the high stacking fault energy of aluminum and its alloys DRV occurs readily and it does not allow the build-up of stored energy by dislocation accumulation inside the material during thermomechanical processes. High stacking fault energy promotes dislocation climb and cross slip which are the basic mechanisms responsible for recovery. DRV leads to rearrangement of dislocations to form subgrains with nearly dislocation free interiors. During DRV, the strain-rate and temperature are expressed by Zener–Hollomon parameter, given by

$$Z = \dot{\varepsilon} \exp\left(\frac{Q}{RT}\right) \tag{2.3}$$

where Q is the activation energy for lattice diffusion and $\dot{\varepsilon}$ is the strain rate. Z determines the subgrain size for a given temperature and strain rate [46]. The equation is generally used for hot working processes where the strain rate and temperature are generally known.

DDRX takes place by the same mechanism as classical recrystallization, i.e., nucleation of new strain free grains and sweeping motion of grain boundaries to create the recrystallized grains. DDRX occurs in metals of medium to low stacking fault energy during thermomechanical process. DDRX occurs when a critical deformation condition is reached and requires large scale high-angle boundary migration. A condition for initiation of DDRX is given by [46]

$$\frac{\rho_m}{\dot{\varepsilon}} > \frac{2\gamma_b}{KLmGb^5} \tag{2.4}$$

where ρ_m is the mobile dislocation density, $\dot{\varepsilon}$ is the strain rate, γ_b is the grain boundary energy, K is a constant fraction of the dislocation line energy that is stored in the newly formed grains, L is mean slip distance of dislocations in these grains, m is the boundary mobility, G is the shear modulus, and b is the Burgers vector. This inequality suggests dependence of DDRX on mobile dislocation density. High rate of recovery occurs in high stacking fault energy materials and reduces ρ_m especially at high strain rates thereby reducing the possibility of DDRX. A critical deformation is necessary to initiate dynamic recrystallization and it originates at high-angle boundaries. Aluminum and its alloys normally do not undergo DDRX because of their high rate of recovery due to high stacking fault energy of Al.

CDRX, on the other hand, takes place by gradual increase in the misorientation between subgrain boundaries. CDRX occurs by the rearrangement of dislocations into subgrain boundaries. Absorption of dislocations, generated during deformation, into these subgrain boundaries increases the misorientation between adjacent subgrains. Several mechanisms of CDRX have been proposed whereby subgrains rotate and achieve a high misorientation angle with little boundary migration. These mechanisms include subgrain growth, lattice rotation associated with sliding [47], and lattice rotation associated with slip [48]. Geometric dynamic recrystallization (GDRX) is somewhat different from DDRX or CDRX in the sense that during deformation, the original grains flatten (compression) or elongate (tension, torsion), and their boundaries become progressively serrated while subgrains form. Ultimately, when the original grain thickness is reduced to about two subgrain sizes, the grain boundaries begin to come in contact with each other locally, causing the grains to pinch-off at the contact points forming new isolated grains [49]. GDRX may be observed during DRV when prior boundary separation becomes equal to subgrain size which in turn is related to Z^{-1}. The boundary separation in the normal direction, d_n, during hot or warm rolling is given by

$$d_n = d_o \exp(-\varepsilon) \tag{2.5}$$

where d_o is the initial grain size and ε is the strain. Low stain rate and stresses lead to small value of Z and onset of GDRX. At high strain rates the DRV stage of deformation will be more prolonged before onset of GDRX.

Jata and Semiatin, were the first to propose CDRX as grain refinement mechanism during FSW of Al-Li alloy [20]. They suggested that high-angle boundaries in the nugget zone are created from low-angle boundaries by means of a continuous rotation of the original low-angle boundaries during FSW. Dislocation glide is shown to be responsible for gradual relative rotation of adjacent subgrains in their model. The strain induces progressive rotation of subgrains with little boundary migration and the subgrains thus gradually transform into high-angle grain boundaries. Su et al. carried out a detailed microstructural investigation of FSW 7050Al [50]. Based on the microstructural analysis and on the basis of occurrence of DRV they suggested CDRX to be the mechanism of dynamic recrystallization in the nugget zone. McNelley et al. suggested that DRV and geometric dynamic recrystallization (GDRX) dominate the microstructure refinement during FSP in the light of high stacking fault energy of aluminum and its alloys [51]. But they could not explain the formation of finer grains of 100 nm in their microstructure. Prangnell and Heason proposed GDRX as grain refinement process driven by grain subdivision using the "stop action" technique during FSW of Al-2195 [52]. Fonda et al. attributed DRV as the grain refinement mechanism based on the fcc shear texture obtained during stop action FSW of Al-Li 2195 alloy and concluded that fine grains form due to simple shear deformation field and there is no occurrence of dynamic recrystallization [53].

All these mechanisms are based on the observation of the final microstructure in the processed/welded material. The evolution path of the fine grains is not defined clearly. Rhodes et al. tried to simulate the FSW condition by tool plunge and extraction method followed by annealing [54]. They reported nucleation (DDRX) and growth to be the grain refinement mechanism during FSP of 7050 Al. Su et al. made an attempt to study the restorative mechanism during FSP in aluminum and its alloys. The authors observed grains with different dislocation densities and in various degree of recovery and proposed that multimechanism of DDRX, DRV, and CDRX act at different stages of microstructure evolution due to heterogeneous plastic deformation. They attributed the formation of 100–400 nm fine grains with high-angle boundaries to discontinuous DRX [11,55].

REFERENCES

[1] W.M. Thomas, E.D. Nicholas, J.C. Needham, M.G. Murch, P. Templesmith, C.J. Dawes, GB Patent No. 9125978.8, Dec 1991.

[2] R.S. Mishra, Z.Y. Ma, Mater. Sci. Eng. R 50 (2005) 1–78.

[3] Z.Y. Ma, Metall. Mater. Trans. A 39 (2008) 642–658.

[4] R.S. Mishra, M.W. Mahoney, S.X. McFadden, N.A. Mara, A.K. Mukherjee, Scripta Mater. 42 (2000) 163–168.

[5] W.B. Lee, S.B. Jung, Mater. Lett. 58 (2004) 1041–1046.

[6] M.S. Moghaddam, R. Parvizi, Mater. Des. 32 (2011) 2749–2755.

[7] S. Mironov, Y.S. Sato, H. Kokawa, Acta Mater. 56 (2008) 2602–2614.

[8] Y.C. Chen, H. Fujii, T. Tsumura, Y. Kitagawa, K. Nakata, K. Ikeuchi, et al., Sci. Technol. Weld. Join. 14 (2009) 197–201.

[9] N. Kumar, M. Komarasamy, P. Nelaturu, Z. Tang, P.K. Liaw, R.S. Mishra, JOM 67 (2015) 1007–1013.

[10] A. Farias, G.F. Batalha, E.F. Prados, R. Magnabosco, S. Delijaicov, Wear 302 (2013) 1327–1333.

[11] J.Q. Su, T.W. Nelson, C.J. Sterling, Philos. Mag. 86 (2006) 1–24.

[12] C.I. Chang, X.H. Du, J.C. Huang, Scripta Mater. 57 (2007) 209–212.

[13] C.I. Chang, X.H. Du, J.C. Huang, Scripta Mater. 59 (2008) 356–359.

[14] Z.Y. Ma, R.S. Mishra, Scripta Mater. 53 (2005) 75–80.

[15] Z.Y. Ma, R.S. Mishra, M.W. Mahoney, Acta Mater. 50 (2002) 4419–4430.

[16] Z.Y. Ma, R.S. Mishra, M.W. Manohey, R. Grimes, Mater. Sci. Eng. A 351 (2003) 148–153.

[17] F.C. Liu, Z.Y. Ma, Scripta Mater. 58 (2008) 667–670.

[18] C.G. Rhodes, M.W. Mahoney, W.H. Bingel, R.A. Spurling, C.C. Bampton, Scripta Mater. 36 (1997) 69–75.

[19] G. Liu, L.E. Murr, C.S. Niou, J.C. McClure, F.R. Vega, Scripta Mater. 37 (1997) 355–361.

[20] K.V. Jata, K.L. Semiatin, Scripta Mater. 43 (2000) 743–749.

[21] A.H. Feng, Z.Y. Ma, Scripta Mater. 56 (2007) 397–400.

[22] W. Woo, H. Choo, M.B. Prime, Z. Feng, B. Clausen, Acta Mater. 56 (2008) 1701–1711.

[23] X.H. Du, B.L. Wu, Trans. Nonferrous Met. Soc. China 18 (2008) 562–565.

[24] Y.S. Sato, M. Urata, H. Kokawa, Metall. Mater. Trans. A 33 (2002) 625–635.

[25] I. Charit, R.S. Mishra, Mater. Sci. Eng. A 359 (2003) 290–296.

[26] D. Yadav, R. Bauri, Mater. Sci. Eng. A 539 (2012) 85–92.

[27] S. Benavides, Y. Li, L.E. Murr, D. Brown, J.C. McClure, Scripta Mater. 41 (1999) 809–815.

[28] J.Q. Su, T.W. Nelson, C.J. Sterling, J. Mater. Res. 18 (2003) 1757–1760.

[29] M.L. Santella, T. Engstrom, D. Storjohann, T.-Y. Pan, Scripta Mater. 53 (2005) 201–206.

[30] W. Yuan, S.K. Panigrahi, R.S. Mishra, Metall. Mater. Trans. A 44 (2013) 3675–3684.

[31] R.S. Mishra, Z.Y. Ma, I. Charit, Mater. Sci. Eng. A 341 (2003) 307–310.

[32] R. Bauri, D. Yadav, G. Suhas, Mater. Sci. Eng. A 528 (2011) 4732–4739.

[33] M. Mahoney, R.S. Mishra, T. Nelson, J. Flintoff, R. Islamgaliev, Y. Hovansky, Friction Stir Welding and Processing, The Minerals, Metals and Materials Society, Warrendale, 2001.

[34] S. Agarwal, C.L. Briant, L.G. Hector Jr., Y.L. Chen, J. Mater. Eng. Perform. 16 (2007) 391–403.

[35] M.W. Mahoney, C.G. Rhodes, J.C. Flintoff, R.A. Spurling, W.H. Bingel, Metall. Mater. Trans. A 29 (1998) 1955–1964.

[36] C.I. Chang, C.J. Lee, J.C. Huang, Scripta Mater. 51 (2004) 509–514.

[37] G. Buffa, J. Hua, R. Shivpuri, L. Fratini, Mater. Sci. Eng. A 419 (2006) 389–396.

[38] R. Nandan, G. Roy, T. DebRoy, Metall. Mater. Trans. A 37 (2006) 1247–1259.

[39] T.U. Seidel, A.P. Reynolds, Metall. Mater. Trans. A 32 (2001) 2879–2884.

[40] Y.S. Sato, H. Kokawa, M. Enomoto, S. Jogan, Metall. Mater. Trans. A 30 (1999) 2429–2437.

[41] Y.J. Kwon, N. Saito, I. Shigematsu, J. Mater. Sci. Lett. 21 (2002) 1473–1476.

[42] L.E. Murr, G. Liu, J.C. McClure, J. Mater. Sci. 33 (1998) 1243–1251.

[43] T. Hashimoto, S. Jyogan, K. Nakata, Y.G. Kim, M. Ushio, FSW joints of high strength aluminium alloys, in: Proceedings of the First International Symposium on Friction Stir Welding, Thousand Oaks, CA, USA, 1999.

[44] Y.J. Chao, X. Qi, W. Tang, Trans. Am. Soc. Mech. Eng. 125 (2003) 138–145.

[45] W.J. Arbegast, P.J. Hartley, Proceedings of the Fifth International Conference on Trends in Welding Research, Pine Mountain, USA, 1998, p. 541.

[46] F.J. Humphreys, M. Hatherly, Recrystallization and Related Annealing Phenomena, second ed., Elsevier, Oxford, 2004.

[47] S.J. Hales, T.R. McNelley, Acta Metall. 36 (1988) 1229–1239.

[48] K. Tsuzaki, X. Huang, T. Maki, Acta Mater. 44 (1996) 4491–4499.

[49] M.E. Kassner, S.R. Barrabes, Mater. Sci. Eng. A 410–411 (2005) 152–155.

[50] J.Q. Su, T.W. Nelson, R.S. Mishra, M.W. Mahoney, Acta Mater. 51 (2003) 713–729.

[51] T.R. McNelley, S. Swaminathan, J.Q. Su, Scripta Mater. 58 (2008) 349–354.

[52] P.B. Prangnell, C.P. Heason, Acta Mater. 53 (2005) 3179–3192.

[53] R.W. Fonda, J.F. Bingert, K.J. Colligan, Scripta Mater. 51 (2004) 243–248.

[54] C.G. Rhodes, M.W. Mahoney, W.H. Bingel, M. Calabrese, Scripta Mater. 48 (2003) 1451–1455.

[55] J.Q. Su, T.W. Nelson, C.J. Sterling, Mater. Sci. Eng. A 405 (2005) 277–286.

CHAPTER 3

Processing Metal Matrix Composite (MMC) by FSP

ABSTRACT

Friction stir processing (FSP) uses the same principle as friction stir welding (FSW) [1]. In FSP a cylindrical tool that has a shoulder and a pin is rotated at high speed while a vertical force plunges the pin into the material. The frictional heat softens the material and as the tool is traversed forward the material is processed and consolidated. The process involves severe deformation and material mixing. The strain during FSP can go up to as high as 40 [2]. The temperature also rises significantly due to the frictional heating and the adiabatic heating arising from the deformation of the material [1]. The material mixing action and the thermomechanical aspect of the process thus offer the possibility of incorporating second phase particles in the stir zone (SZ) and make a composite.

3.1 PROCESSING APPROACHES

Mishra et al. were the first to report fabrication of Al–SiC composite by FSP [3]. Based on that work several research studies were taken up to fabricate surface or bulk composites by FSP. These studies used different approaches of incorporating the ceramic particles into the matrix material. These approaches are described in the following sections.

3.1.1 Groove Filling Method

In the original work, Mishra et al. first applied a layer of SiC powder on the 5083 Al plate and carried out FSP over the dried layer to incorporate the SiC particles into the SZ. Morisada et al. later fabricated multiwall carbon nanotube (MWCNT) reinforced AZ31 composite by FSP. In their approach, a groove, 1 mm in width and 2 mm in depth (1 mm × 2 mm) was first cut on the AZ31 plate and it was filled with MWCNTs. FSP was then carried out along the groove at a rotation speed of 1500 rpm with traverse speeds varying from 25 to 100 mm/min [4]. They also processed AZ31–SiC and 5083 Al–Fullerene

Metal Matrix Composites by Friction Stir Processing. DOI: http://dx.doi.org/10.1016/B978-0-12-813729-1.00003-6

composites using the same method [5,6]. The groove filling method, which was used later by many other researchers, is shown schematically in Fig. 3.1 [7]. Fig. 3.2 shows the SEM images of the AZ31−SiC composite processed by this method [5]. The groove filling method gained considerable popularity to fabricate a variety of metal matrix composites (MMCs) by FSP [7−15]. Lee et al. made deeper grooves (6 mm) on AZ61 Mg plates and incorporated SiO_2 particles to make bulk composites by FSP [7,8]. Fig. 3.3A shows an SEM image of the AZ61-5 vol.% SiO_2 composite processed by a single FSP pass. Sun et al. also reported fabrication of bulk Mg−nano-SiC composite by the groove method [9]. Often the groove was first closed with a blunt tool (without pin) before processing as shown in Fig. 3.1B. Closing the groove with an additional sheet as top cover on the original plate has also been reported [11]. Both single pass and multipass FSP processes have been applied to make composites by the groove filling method. The method is fairly simple and generally yields composites with

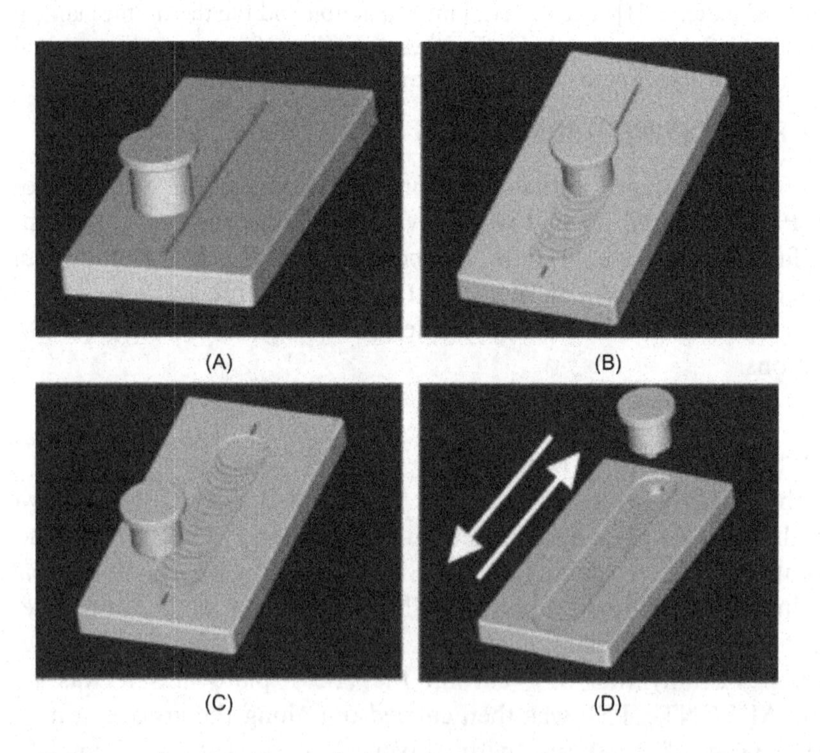

Figure 3.1 Schematic of composite fabrication by friction stir processing: (A) cutting groove and inserting particles, (B) using a flat tool to close the groove, (C) applying a tool with a fixed pin to undertake the FSP, and (D) conducting multiple FSP passes [7].

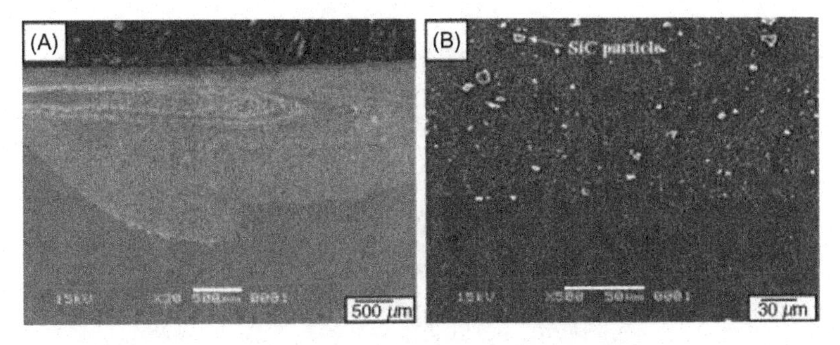

Figure 3.2 SEM images of (A) stir zone with the SiC particles and (B) interface zone between the AZ31/SiC composite and the AZ31 [5].

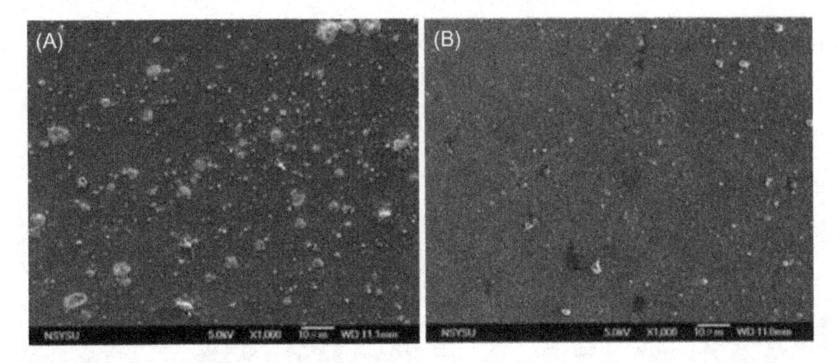

Figure 3.3 SEM images of AZ61-5 vol.% SiO_2 composite fabricated by (A) single pass and (B) multiple FSP passes (four passes) [7].

uniform particle distribution. Moreover, the uniformity of particle distribution improved with number of FSP passes as shown in Fig. 3.3B for the AZ61-5 vol.% SiO_2 composite [7].

3.1.2 Drill-Hole Method

In a similar approach a series of holes made on the plate have been used to incorporate the reinforcement particles. The method is shown schematically in Fig. 3.4. The holes contained within the shoulder diameter are filled with the reinforcement powder and FSP is carried out over them. Dixit et al. processed Nitinol (NiTi) reinforced Al 1100 composite by the drill-hole method [16]. They drilled four small holes, 1.6 mm in diameter and 76 mm deep on the Al 1100 plate at about 0.9 mm below the surface. The holes were drilled in such a manner that they are distributed within less than the shoulder width so that all the holes are processed during a single pass of FSP. The NiTi particles were packed in the holes and FSP was carried out over them at a tool

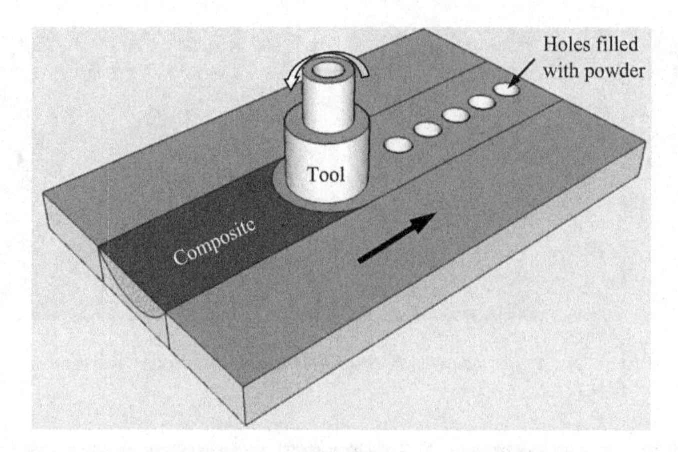

Figure 3.4 Schematic of the drill-hole method.

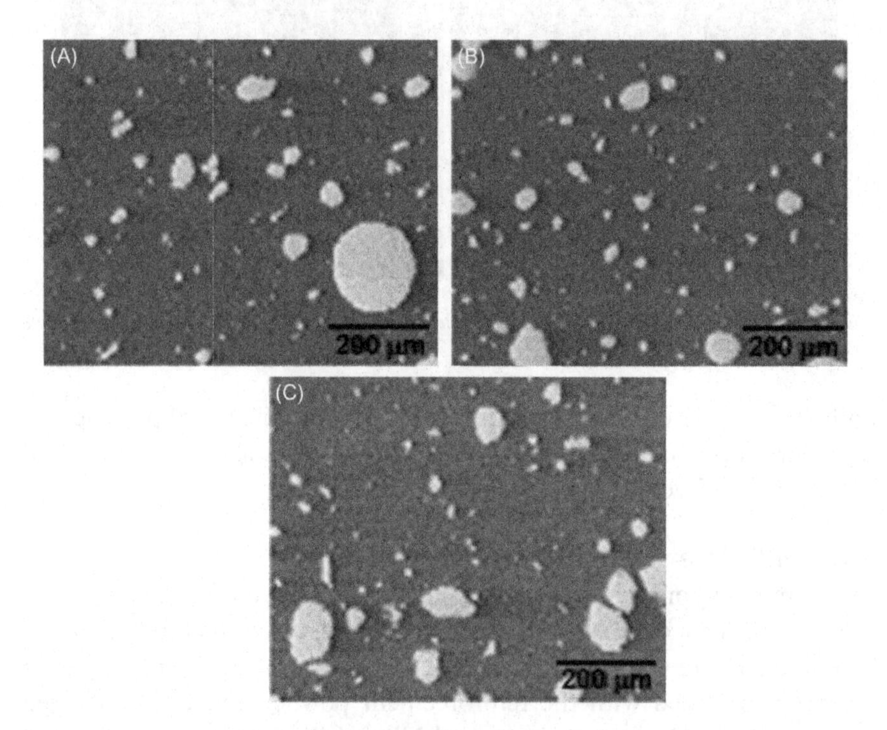

Figure 3.5 SEM images showing uniformly distributed NiTi particles in various parts of the nugget region of FSP composites (A)–(C) [16].

rotation speed of 1000 rpm and traverse speed of 25 mm/min. The process resulted in uniform distribution of NiTi particles in the Al matrix as shown in Fig. 3.5. 6061 Al composites reinforced with NiTi particle was also fabricated by the same method using multiple passes of FSP [17]. In this case also a uniform distribution of NiTi particles was

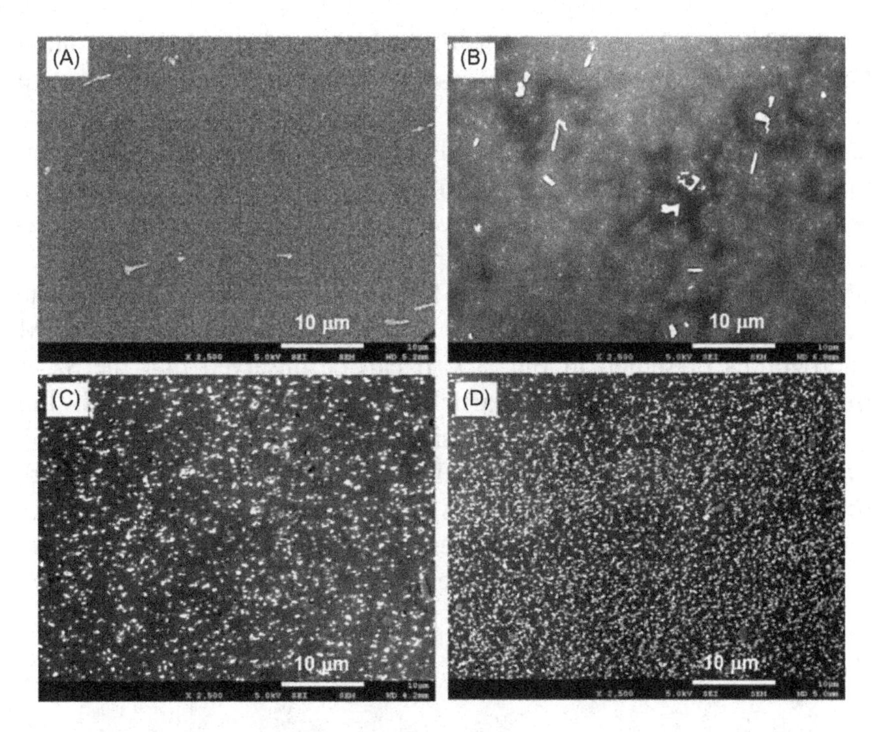

Figure 3.6 SEM images of (A) FSPed Al, (B) FSPed Al–CNTs composite, (C) FSPed Al-Al_2O_3 composite, and (D) FSPed Al–Al_2O_3–CNTs composite [19].

obtained. Liu et al. incorporated up to 6 vol.% MWCNT in Al by the drill-hole method and reported good bonding between the nanotubes and the Al matrix [18]. Du et al. later incorporated nanosize Al_2O_3 particles along with MWCNT to fabricate Al–Al_2O_3–CNT hybrid composite [19]. Fig. 3.6 shows SEM images of the Al–CNTs, Al–Al_2O_3, and Al–Al_2O_3–CNTs composites. The drill-hole approach has been also used to incorporate TiO_2 nanofibers in Al [20]. Further, it has been also utilized to make Al–TiC functionally graded composites by a friction stir additive manufacturing process [21]. It may be noted that though a variety of MMCs have been fabricated using the drill-hole method the number, design and size of the holes may vary from one work to another.

3.1.3 Powder Metallurgy Route

In certain cases a combination of powder metallurgy (PM) and FSP has been utilized to process MMCs. In this approach, first a small billet containing the desired reinforcement in the matrix is made by conventional PM process. It involves mixing and blending of matrix and reinforcement powders followed by compaction and sintering. The sintered billet is then subjected to FSP which results in final consolidation. The initial sintering in this case is utilized just to provide enough strength to the billet so that it does not fracture during FSP. The final densification takes place during FSP and the redistribution of the reinforcement also occurs due to the material flow in FSP. Therefore a fully dense composite with all leftover porosity closed and uniform distribution of the reinforcement is expected through this combination of PM and FSP. Lee et al. fabricated Si particle reinforced Al composite by this approach [22]. The effect of FSP can be seen in the SEM micrographs presented in Fig. 3.7. It can be observed that the SZ is free of porosity and the Si particles are uniformly distributed. Liu et al. fabricated Al–CNTs composite by the same approach [23–25]. They also applied hot rolling as a final step after FSP to develop high performance Al–CNTs composites [24]. FSP redistributed the CNTs uniformly and the hot rolling process was utilized to align them along the rolling direction. The entire scheme is shown in Fig. 3.8.

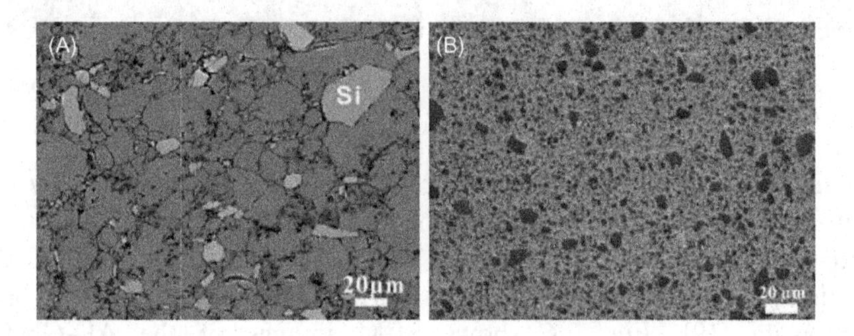

Figure 3.7 SEM images of (A) Al-10 Si sintered at 803 K and (B) Al-30 Si produced by four FSP passes [22].

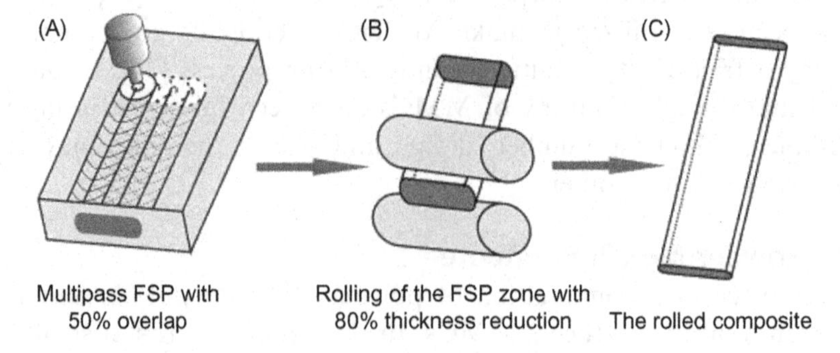

Figure 3.8 Schematic of CNT/2009 Al composite fabrication. Multipass FSP (A) is followed by hot rolling (B) to make the CNTs aligned in the final composite (C).

3.1.4 In situ Method

Generating the reinforcement particles in situ during the process instead of adding them separately has many advantages such as clean particle–matrix interface, better particle–matrix bonding, and thermodynamically stable and finer particles. The synergistic effect of frictional heating and the exothermic nature of the in situ reaction has been utilized to process in situ composites by FSP. Two approaches have been generally used to initiate the in situ reaction. In one approach, the precursor powder is first loaded in the base plate through grooves or drilled holes and the assembly is then subjected to FSP. The constituent powders react with themselves and/or with the Al matrix during FSP to generate the reinforcement particles during the process. The exothermic nature of the reaction makes it self-sustaining once it initiates by the frictional heat. Qian et al. produced Al–Al_3Ni composite using this approach by filling Ni powder into the groove and subjecting it to FSP [26]. Ni reacted with Al during FSP and generated Al_3Ni particles in situ. Ke et al. fabricated similar type of Al–Al_3Ni composite but in their case the Ni powder was filled in drilled holes [27]. The method has been even utilized to generate polymer derived ceramics (PDC) into a Cu matrix [28,29]. The process for making the PDC composite involved incorporation of a polymer precursor by FSP through the groove route followed by heat treatment to pyrolize the polymer into ceramic and subjecting it to FSP again.

The other approach of the in situ technique utilizes PM. The precursor powders are mixed with the matrix powder and consolidated by PM. The PM billet is then subjected to FSP to process the in situ composite. Al–Al_3Ti–Al_2O_3 composites were produced through this approach by subjecting an Al–TiO_2 billet made by PM to reactive FSP. The constituents, Al and TiO_2, reacted during FSP and produced fine Al_3Ti and Al_2O_3 particles in the Al matrix [30,31]. Hsu et al. processed Al–Al_3Ti nanocomposites by a similar approach [32]. The reactive FSP approach has been followed to process a variety of Al based in situ composites [33–36].

3.2 EFFECT OF PROCESS PARAMETERS AND TOOL GEOMETRY

FSP involves complex material flow and plastic deformation. The process parameters and tool geometry have a significant effect on the material flow pattern, temperature, strain and strain rate thereby influencing the microstructure and particle distribution in composites fabricated by FSP. The tool related parameters essentially are the geometry of the tool shoulder and pin and the different features on them. The following discussions on these parameters will help their selection for processing defect free composites by FSP.

3.2.1 Process Parameters

The frictional heat at the tool–workpiece interface softens the material which then flows around the pin as the rotating tool traverses forward. The amount of heat input into the material depends on the ratio of tool rotation speed (ω) to traverse speed (v), i.e., the ω/v ratio. As the rotational speed is increased more frictional heat is generated and the deformation also becomes more intense generating higher adiabatic heating. On the other hand, as the traverse speed is decreased the tool stays for longer time in a given area and consequently that volume of material experiences more heat. Therefore higher the ω/v ratio, higher is the heat input into the material. The maximum temperature in the SZ can be estimated from the relationship given by Arbegast and Hartley as shown in Eq. 2.2 in Chapter 2, Introduction to Friction Stir Processing (FSP) [37]. The maximum temperature during FSW/FSP of various aluminum alloys is found to be in the range $0.6–0.9T_m$ where T_m is the melting point.

As stated earlier, FSP also involves intense plastic deformation. The strain during FSP has been estimated to be as high as 40 [2]. Buffa et al. theoretically showed that the effective strain distribution was not symmetric about the weld line during FSW and maximum strain existed on the advancing side [38]. Eq. 2.1 gives the strain rate which was estimated considering a torsion type deformation by Chang et al. [39]. A maximum strain rate of 160/s has been estimated in FSW and this confirms the existence of intense plastic deformation during FSW/FSP.

While processing composites by FSP, Lim et al. varied the rotational speeds from 1500–2500 rpm and investigated its effect on the distribution on CNTs in the Al matrix [11]. They concluded that the distribution of the nanotubes improved with increasing rotation speed. Similar observations were made by a number of authors for a variety of MMCs reinforced with different particles by FSP [12,40–43]. All these studies have shown that the particle distribution generally improves with increasing rotation speed. An example of the effect of

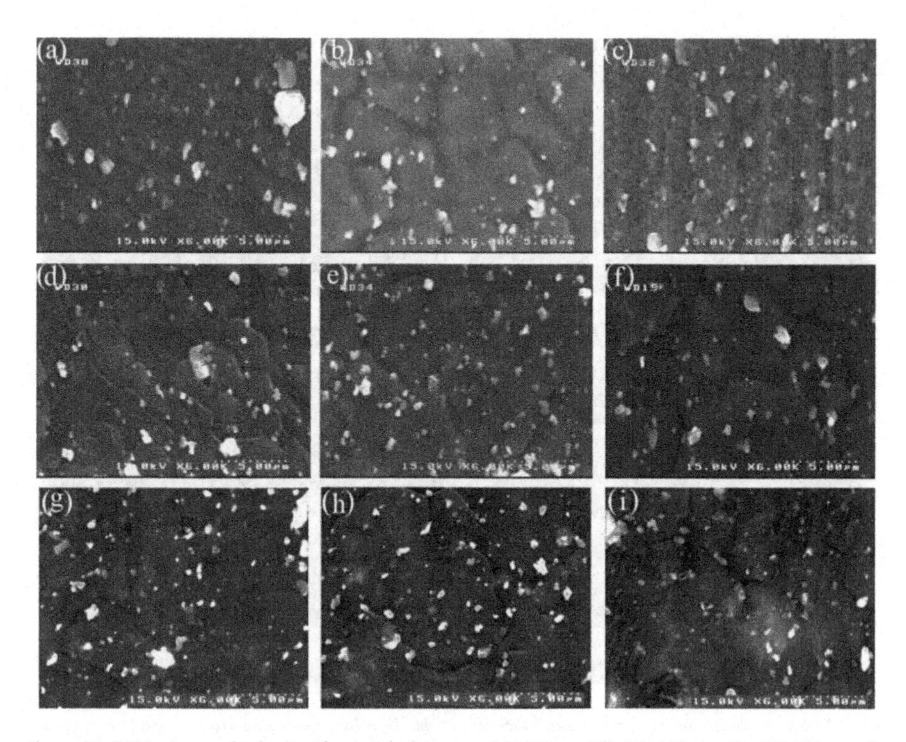

Figure 3.9 SEM micrographs showing the particle dispersion: (A) 800 rpm-2P; (B) 800 rpm-3P; (C) 800 rpm-4P; (D) 1000 rpm-2P; (E) 1000 rpm-3P; (F) 1000 rpm-4P; (G) 1200 rpm-2P; (H) 1200 rpm-3P; and (I) 1200 rpm-4P [41].

rotation speed on particle distribution is shown in Fig. 3.9 [41] for AZ31/Al_2O_3 nanocomposites. The cluster size of alumina particles reduced after every FSP pass and the homogeneity of particle distribution improved after multipass FSP [41]. The authors also reported the formation of onion ring patterns in the samples processed with 1000 and 1200 rpm and absence of the same in 800 rpm samples. The distance between onion rings is equal to the ratio of forward motion to rotation speed. These bands or rings formed due to material flow from the warmer zone in the top to cooler zone below [6]. At lower rotation speed the heat input is not enough and the onion ring could not form due to less material flow. The traverse speed is also an important parameter to consider when it comes to particle distribution in composites made by FSP. For example, Farazi and Asadi reported that distribution of Al_2O_3 particles in AZ31 Mg improved when the traverse speed was decreased from 80 mm/min to 40 mm/min at 900 rpm [43]. This was attributed to increasing heat input as the ω/v ratio increased with decreasing traverse speed (v). The higher heat input causes greater

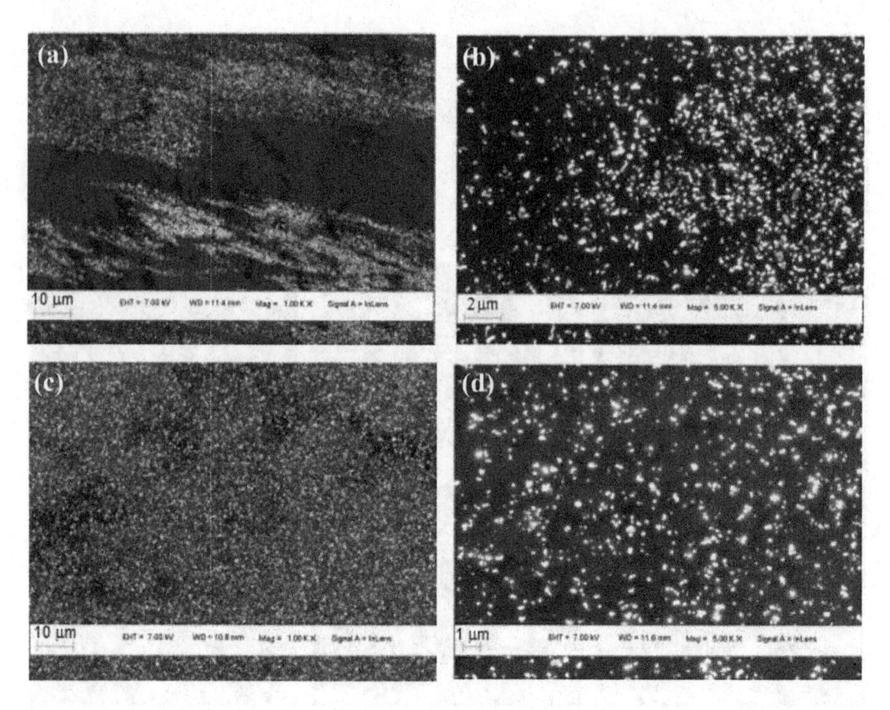

Figure 3.10 SEM images of the transverse cross-sections of the Al–Al_2O_3 composites produced by FSP. (A) and (B) represent the two FSP passes condition, while (C) and (D) represent the four FSP passes condition [44].

softening of the material and it flows easily around the pin thus giving rise to better stirring action. It has also been reported in many cases, especially for fine particles, that a single pass of FSP may not be sufficient to homogeneously distribute the reinforcements [11]. Hence, a majority of the studies have employed multipass FSP to disperse the reinforcement into the matrix [7,9,12,15,17,41,43–45]. The effect of two and four FSP passes on the distribution of Al_2O_3 particles in Al matrix can be seen in Fig. 3.10 [44]. Four FSP passes resulted in significant improvement in particle distribution [44]. It has also been suggested that changing the tool rotation direction between passes can have a balancing effect on the material flow as the advancing side in one pass becomes the retreating side in the next and this yields a more uniform particle distribution [12].

3.2.2 Tool Geometry

The tool geometry affects heat generation and the force and torque experienced by the material and thus has a great influence on the deformation and material flow around the tool. The important parameters with regard to the tool geometry are shoulder diameter, profile and surface angle and pin geometry including pin size, shape, and profile.

The friction between the shoulder and the workpiece generates most of the heat and hence, the shoulder diameter is important. The shoulder also confines the plasticized material for it to be processed. Generally a concave shoulder surface gives rise to a sound weld. Moreover, various shoulder profiles have been designed to suit a range of materials and processing conditions. Different shoulder profiles such as scroll, spiral, fillet, and cavity have been used for achieving better weld properties [1,46–49]. With the help of the special features these profiles improve the coupling between the shoulder and the workpiece by preventing the plasticized material from being expelled from beneath the shoulder.

The pin geometry that includes the size, shape, and pin profiles significantly influences the material flow. Commonly used pin shapes are cylindrical, triangular, three flat, trivex, and conical [50]. Square, pentagonal, and hexagonal cross-sections have been also explored in recent time [51]. Complex pin profiles such as threads, whorl, and triflute have been introduced to achieve better material flow and mixing [1]. A trifluted tool pin results in more material flow compared to a simple cylindrical pin [1,46,50]. The correlation between pin geometry and material flow can be characterized by the ratio of dynamic to static volume, i.e., the ratio of volume of the material swept by the pin to the volume of the pin itself. This ratio is 1.1:1 for conventional cylindrical pins [1]. Pins with special features exhibit a higher ratio. For example, the whorl and triflute pin profiles have a ratio of 1.8:1 and 2.6:1, respectively, for similar root diameter and pin length [1].

Bahrami et al. processed Al 7075/SiC composites using five different pin geometries, i.e., threaded tapered, triangular, square, four-flute square, and four-flute cylindrical and found that the particle distribution was more uniform with the threaded tapered tool [52]. Azizieh et al. fabricated AZ31/Al_2O_3 composites using plain and threaded cylindrical tool pin [41]. They reported an almost uniform distribution of Al_2O_3 particles with the threaded tool whereas the plain tool led to clustering of particles. The better distribution with the threaded tool was attributed to the improvement in the material flow due to the downward movement of material along the threads. A better distribution of nano-Al_2O_3 particles in AZ91 was reported for square pin compared to a cylindrical pin with the same processing parameters [43]. This was attributed to a pulsating stirring action that a square pin produced. The square pin also produces a higher dynamic to static volume ratio (1.56) compared to simple cylindrical pins [53]. Sahraeinejad et al. reported that particle distribution was nonuniform with different particle concentrations in upper and lower part of the SZ in 5059 Al based composite processed with a 3-flat pin tool [12]. This was ascribed to limited material flow vertically with a 3-flat tool. The particle distribution was improved significantly by utilizing a simple threaded tool in combination with the 3-flat tool. The authors argued that the threaded tool provided sufficient heat input to ensure material flow through which the particles could easily move around the tool [12].

3.3 MODIFICATION OF CAST MMCs BY FSP

Discontinuously reinforced aluminum (DRA) matrix composites often exhibit nonuniform particle distribution. The extent of this microstructural heterogeneity depends on several factors such as particle size and shape, alloy composition, and solidification conditions. The clustering or heterogeneity of particle distribution adversely affects the strength [54,55], ductility [56,57], fatigue [58], damage evolution [59,60], and fracture behavior [61,62] of DRA composites. Therefore it is imperative to develop processing techniques which can minimize or eliminate the microstructural heterogeneity in these composites. FSP can be a promising technique for this purpose due to the extensive material flow and mixing action that occur during the process.

Berbon et al. first reported microstructure homogenization of DRA composite by FSP [63]. Al composites reinforced with fine intermetallic particles were processed by cryomilling and hot isostatic pressing (HIP). The composite exhibited inhomogeneous microstructure with Al_3Ti intermetallic particles segregated in certain areas. Subjecting the composite to extrusion could not homogenize the microstructure and as a result the composite suffered from low ductility. FSP of the HIPed composite, on the other hand, dramatically improved the homogeneity of particle distribution and the composites consequently exhibited much higher ductility. The technique was later applied for cast composites to homogenize the microstructure. Yadav and Bauri utilized FSP as a secondary processing tool to homogenize the microstructure of Al–TiB_2 and Al–TiC in situ composites processed by the melt-reaction route [64,65]. The in situ formed particles (TiC/TiB_2) were segregated along the interdendritic regions (grain boundaries) in the cast composites (Fig. 3.11A). As shown in Fig. 3.11B, the particle distribution greatly improved with just a single pass of FSP. The stark difference in the particle distribution between the as-cast and friction stir processed (FSPed) composite can be clearly seen in Fig. 3.11C which shows the interface between these two regions. The microstructure was completely homogenized after two FSP passes (Fig. 3.11D). The casting defects were also eliminated after the FSP passes. The homogenization and the grain refinement imparted by FSP improved both strength and ductility of the composites. Similar observations were made for numerous other Al based cast composites such as A339/SiC_p, A356/Al_3Ti, AA2024/Al_2O_3, AA7075/TiB_2, and 6061 Al/ZrB_2 which were subjected to FSP [66–70]. In every case the

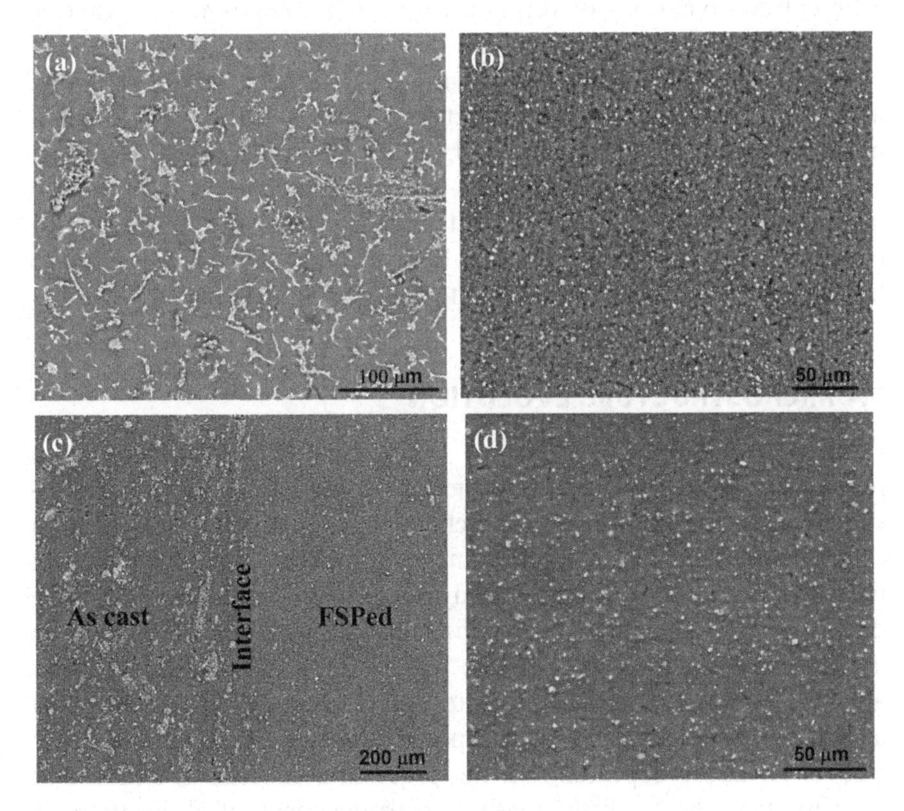

Figure 3.11 SEM images of Al–TiB_2 in situ composite (A) as-cast and (B) after single pass FSP. (C) shows the interface between the as-cast and stir zone, and (D) microstructure of the composite after two FSP passes.

mechanical properties of the composites improved significantly after FSP owing to homogenization of the microstructure and grain refinement.

It is worth mentioning here that the processing parameters, namely the rotation and traverse speeds, have to be optimized to obtain a defect free SZ in the processed composites [71]. It may also be noted that parameters which yield a defect free SZ in the unreinforced alloy may not be sufficient to process the composite as the material flow in the matrix is partially hindered by the particles and the composite is also stiffer [72]. The amount of heat input into the material is directly proportional to the ratio of rotation to traverse speed. For a particular material a certain ratio is needed below which the heat input may not be sufficient to plasticize material for it to flow around the tool leading to defects in the SZ. Bauri optimized the process parameters for FSP of Al–TiC cast in situ composites by carrying out FSP with different combinations of rotation (640, 800, 1000 rpm) and traverse speeds (60, 120, 150 mm/min) [71]. It was found that lower rotational speeds (640 and 800 rpm) with all the three traverse speeds (60, 120, and 150 mm/min) give rise to defects in the SZ and the particle distribution was also not homogenized. A rotational speed of 1000 rpm and traverse speed of 60 mm/min was reported to be optimum to homogenize the composite without any defects in the SZ. With the right set of process parameters the FSP technique can be also used to homogenize MMCs processed by other methods such as PM [73,74].

3.4 MICROSTRUCTURE EVOLUTION

Friction stir processing is known to be an effective tool for grain refinement. As the material is subjected to thermomechanical processing during FSP the grains are refined by dynamic recrystallization (DRX) process [1,75–77]. DRX in Al alloys can occur by different mechanisms such as continuous dynamic recrystallization (CDRX), discontinuous dynamic recrystallization (DDRX), and geometric dynamic recrystallization (GDRX) [78,79]. Dynamic recovery (DRV) also plays an important role in restoration and grain refinement of Al alloys. All these mechanisms are reported to be operating during FSW/FSP of Al alloys depending on the processing conditions, strain, and temperature. DRV which involves rearrangement of dislocation in low-energy configuration can readily occur in Al due to its high

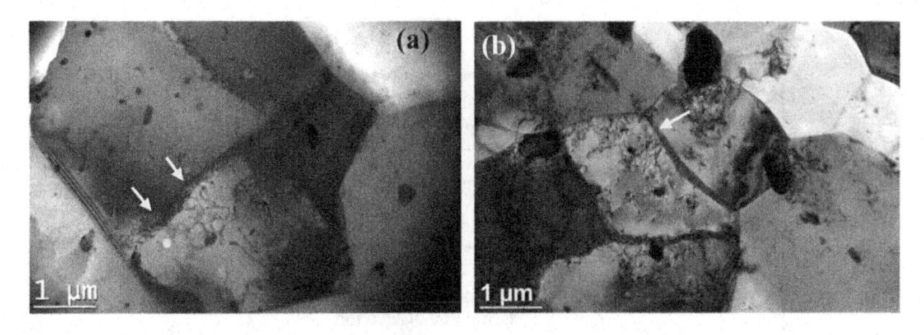

Figure 3.12 TEM micrographs of FSPed Al showing (A) dislocation arrangement in subgrain boundaries as indicated by the arrows, (B) diffraction contrast (light and dark contrast) across subgrains.

stacking fault energy (~200 mJ/m^2). This is consistent with the observations of CDRX and GDRX that involves dislocation rearrangement into subgrain boundaries [76,80]. CDRX takes place by gradual increase in the misorientation between the subgrains. Dislocations generated during the high temperature deformation are absorbed in the subgrain boundaries and progressively increase the misorientation between them and thus turning them into grain boundaries. Fig. 3.12 shows TEM images of an FSPed Al sample. The TEM micrograph in Fig. 3.12A shows that the subgrain boundary is composed of an array of dislocations and dislocations are being absorbed into the subgrain boundaries (arrows). The diffraction contrast between the subgrains observed in Fig. 3.12B confirms that the misorientation across them has increased. DDRX, on the other hand, takes place by the classical recrystallization mechanism and requires a high density of free dislocations, i.e., accumulation of dislocations or buildup of strain which can occur only in highly strained regions. The overall scenario of microstructure evolution in Al alloys during FSW/FSP is shown schematically in Fig. 3.13.

In the composites fabricated by FSP the overall mechanism of grain refinement remains the same except that the presence of particles may have some effect on the nucleation and grain growth. Guo et al. studied the effect of nano-Al_2O_3 particles on the grain structure of AA6061 composites fabricated by FSP [44,81]. It was reported that the grain size of the AA6061/nano-Al_2O_3 composite was finer (2.5 μm) compared to the unreinforced FSPed alloy (5.9 μm). Similar observations on grain refining effect of nanoparticles in DRA MMCs were made by Khodabakhshi et al. [82] and Du et al. [19]. The finer grain size in the composites was attributed to pinning effect of the particles on the grain

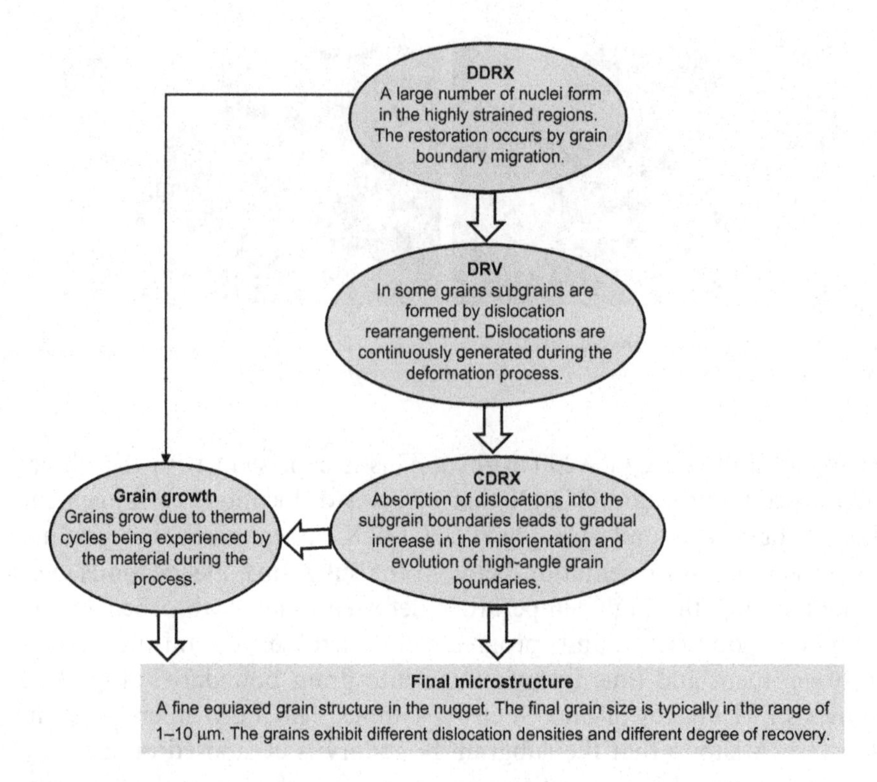

Figure 3.13 Schematic of microstructure evolution during FSW/FSP of Al alloys.

boundaries that restricts the grain growth. The Zener pinning effect of dispersed particles on the grain growth rate during recrystallization of metals can be described by the following equation [79].

$$\frac{dR}{dt} = m(P - P_z) = \left(\frac{\alpha\gamma_b}{R} - \frac{3F_V\gamma_b}{2r}\right) \quad (3.1)$$

where m is the boundary mobility, γ_b is the boundary energy, P is the driving pressure due to the curvature of the grain boundaries, P_z is the Zener pinning pressure, R is the radius of the grain, r is the radius of the particles, F_V is the volume fraction of particles, and α is a geometric constant.

As can be seen in Eq. (3.1), grain growth stops $(dR/dt = 0)$ when $P = P_z$. The Zener limiting grain size is obtained at $\alpha = 1$ when the mean grain radius equals the radius of curvature (R) of the grain boundaries.

Particle stimulated nucleation (PSN) can also occur during recrystallization of composites and this enhances the nucleation rate and thus refines the grain size of the matrix. However, PSN can only occur for particles larger than a critical size (>1 μm) and at lower processing temperature and higher strain rate that help dislocation accumulation near the particles [79].

The microstructure development in FSPed polymer derived ceramic (PDC) particle reinforced Cu composites was studied by electron backscatter diffraction (EBSD) analysis [29]. Fig. 3.14 shows the EBSD (grain boundary + IPF) images of the FSPed Cu and the composite processed under same conditions. The composite exhibited a much finer grain size (1–3 μm) which was lower by more than one order of magnitude compared to the unreinforced FSPed Cu (100 μm) owing to the pinning effect of finely dispersed particles. The corresponding grain size distributions are also shown in Fig. 3.14C and D. The microstructure of the composite is characterized by equiaxed fine grains (average grain size ~2 μm) with a narrow grain size distribution. The composite

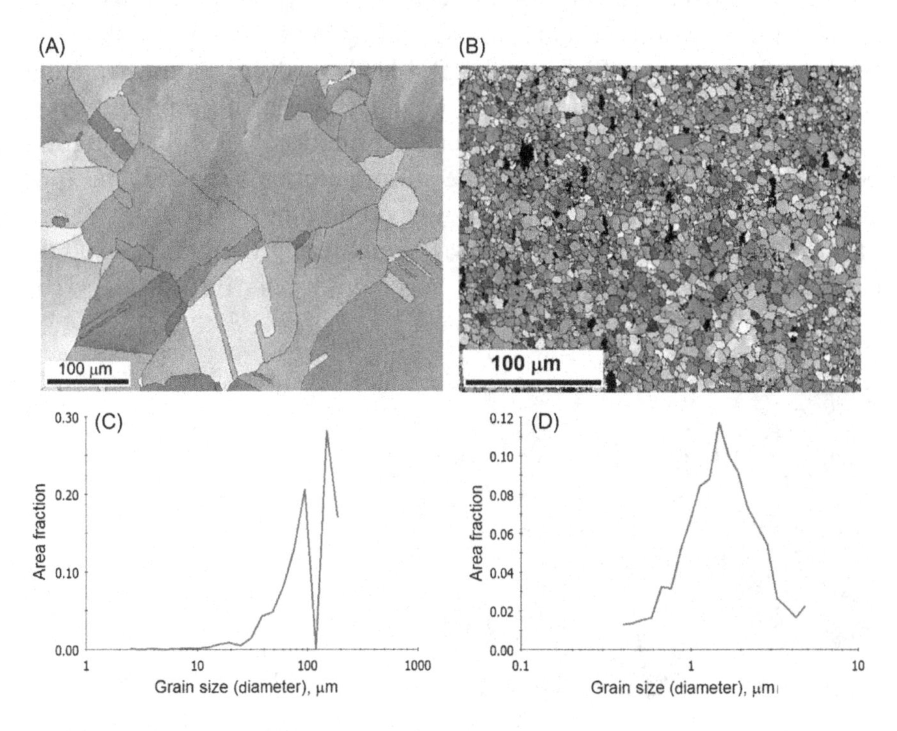

Figure 3.14 EBSD (grain boundary + IPF) maps of (A) FSPed Cu and (B) Cu–PDC composite. (C) and (D) The corresponding grain size distributions.

also exhibited a high fraction (>80%) of high-angle grain boundaries (HAGBs). A high fraction of HAGBs and narrow grain size distribution has been reported in Al–Mg/TiO_2 composites as well [82]. Additionally, it was reported that the mean misorientation angle increased with increasing volume fraction of TiO_2 particles. The strong pinning effect of the fine PDC particles in the Cu composite was demonstrated by analyzing the grain structure in the cross section at the retreating side. A bimodal grain size distribution with particle-rich areas showing finer grains (2 μm) and particle-lean regions showing coarser grains (10 μm) was observed (Fig. 3.15). The fine grained microstructure developed in FSP is often prone to abnormal grain growth (AGG) when subjected to post-processing thermal cycles, e.g., solution heat treatment as the pinning parameter drops drastically at high temperature [83]. The strong pinning effect of the uniformly dispersed fine particles has shown to even prevent the AGG in MMCs subjected to high temperature annealing after FSP [81,84].

The restoration mechanism in DRA MMCs processed by FSP is not different from that in unreinforced FSPed Al alloys. As described earlier, DRV readily occurs in Al during FSP owing to its high stacking fault energy (SFE) [77]. DRV leads to dislocations arranging themselves into low-angle subgrain boundaries. The absorption of dislocations, generated during further deformation, into the subgrain boundaries gradually increases the misorientation across them thus creating a conducive environment for occurrence of CDRX. Yadav and Bauri, for example, have shown by analyzing the grain boundary character that grain refinement mechanism in FSPed Al–TiB_2

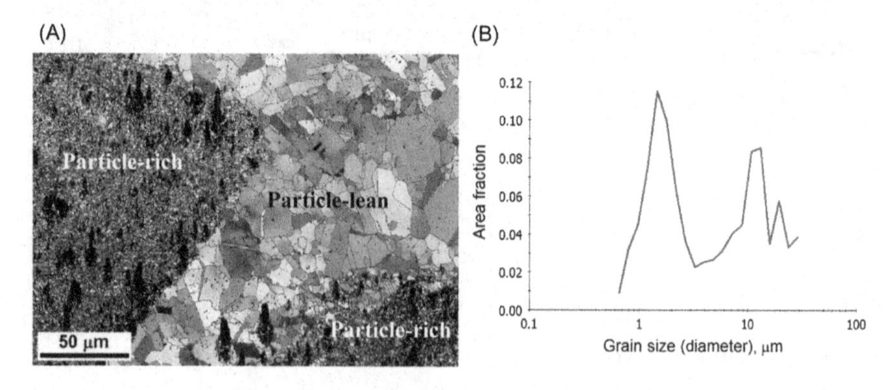

Figure 3.15 (A) EBSD map and (B) grain size distribution of the Cu–PDC composite showing a bimodal grain size in the cross section at the retreating side.

composite resembles CDRX [64]. Fig. 3.16 shows the grain boundary map superimposed on the image quality map of the Al–TiB_2 composite matrix. Different types of boundaries, namely, subgrain boundaries (2–5 degrees), low-angle (5–15 degrees) and high-angle grain boundaries (>15 degrees), are shown in different shades. The presence of the mixed orientation i.e., subgrain + low-angle (white arrows) and low-angle + high-angle character (black arrows) in the same boundary indicates gradual transformation of subgrain boundaries to low-angle and then to HAGBs, a typical characteristic of CDRX. Sharma et al. have also reported occurrence of CDRX in Al–TiC composites processed by FSP [21]. In a low SFE material like Cu (SFE ~78 mJ/m^2) dislocation climb, cross slip and annihilation is relatively difficult and hence, dislocation density can build up. This creates a favorable condition for occurrence of DDRX as was reported by Kumar et al. for Cu–PDC composites [29]. Additional dislocations generated due to the differential flow behavior of the stronger ceramic particles and the softer Cu matrix also contributes to increase the dislocations density which favors DDRX. It may be, however, noted that DDRX can take place during FSP of Al and its composites as well depending on the strain and temperature experienced by the material [75].

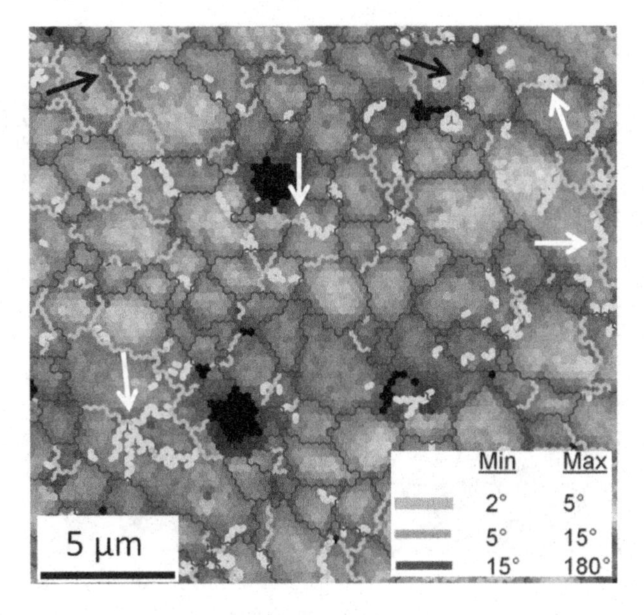

Figure 3.16 Grain boundary map superimposed on image quality map of FSPed Al–TiB_2 composite. White arrows indicate mixed orientation of subgrain + low-angle boundaries and black arrows indicate low-angle + high-angle boundary mixed character.

3.5 MECHANICAL PROPERTIES

MMCs fabricated by FSP generally show better mechanical properties compared to the base and FSPed alloy. The improvement in the hardness and strength has been ascribed to the combined effect of grain refinement and strengthening from the reinforcement. Liu et al. measured the hardness across the SZ as a function of volume fraction of CNTs in Al/MWCNT composites [18]. The hardness improved significantly with addition of CNTs and the improvement increased with increasing CNT content as shown in in Fig. 3.17. The typical stress–strain curves of the composites are shown in Fig. 3.18. The strength also increased with increasing volume fraction of MWCNTs. Similar trends in hardness and strength have been reported for a variety of particle reinforced composites fabricated by FSP [9,12,19,22,26,29–33,44]. The ductility of composites fabricated by FSP is generally found be better than the conventional MMCs and hence, a better strength–ductility combination can be obtained in the MMCs processed by FSP. The improved ductility can be attributed to a clean particle–matrix interface free of any deleterious reaction product, good interfacial bonding, and uniform particle distribution.

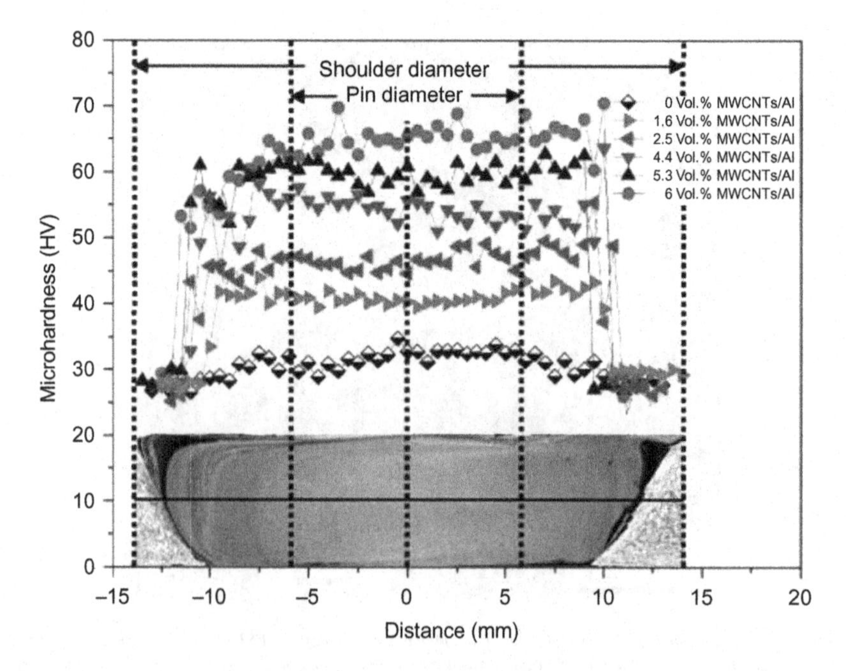

Figure 3.17 Hardness profile of Al/MWCNTs composites as a function of CNT volume fraction [18].

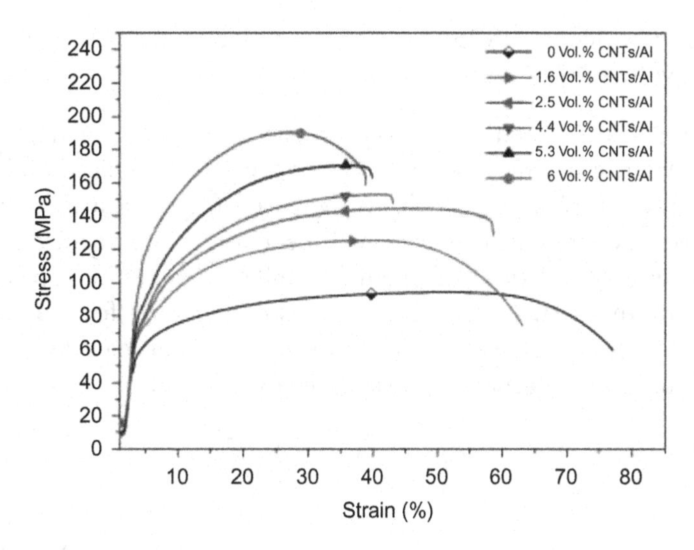

Figure 3.18 Tensile stress–strain curves of Al/MWCNTs composites fabricated by FSP [18].

The strengthening mechanisms in the composites processed by FSP remain the same as those in conventional MMCs. The main strengthening mechanisms in particle reinforced MMCs are: (i) Orowan strengthening, (ii) grain and substructure strengthening, (iii) quench hardening resulting from the dislocation generated due to CTE mismatch between the matrix and the particles, and (iv) work hardening due to the misfit strain between elastic particles and plastic matrix [85]. The major contributors to strengthening in MMCs fabricated by FSP are Orowan mechanism due to the dispersion of finely divided particles and fine grain size. As is well known, the contribution from fine grain size can be derived from the Hall–Petch relationship:

$$\sigma_y = \sigma_i + \frac{k}{\sqrt{d}} \tag{3.2}$$

where σ_i is the *friction stress* that represents the overall resistance of the lattice to dislocation movement, k is known as *locking parameter* which is a measure of relative hardening contribution of grain boundaries, and d is the grain diameter. For pure Al, $\sigma_i = 13$ MPa and $k = 74$ MPa $\mu m^{1/2}$ [86]. The value of the Hall–Petch slope (k) may be higher for severely deformed ultrafine grains due to high dislocation density. The material experiences severe plastic deformation and goes through dynamic recrystallization during FSP. Depending on the dislocation density in the interior of the fine grains and how it affects the slope the k value in FSPed Al may be different from the previously reported values [87].

The Orowan strengthening due to particle dispersion depends on the interparticle distance (λ). As the particle size is reduced and/or the volume fraction (F_V) of particles is increased the interparticle distance decreases. The strength increases with decreasing interparticle spacing. There have been various refinements to the expression of λ in order to incorporate the particle size effect and also to account for the randomness of obstacle distribution. Considering the particles as spherical with radius r, the interparticle spacing is given by [88].

$$\lambda = \left[\left(\sqrt{\frac{\pi}{F_V}}\right) - 2\right]\left(\sqrt{\frac{2}{3}}\right) r \tag{3.3}$$

The Orowan strengthening contribution can be obtained from the following equation [88]:

$$\sigma_O = M \frac{0.81 G_m b}{2\pi(1-\nu)^{1/2}\lambda} \ln\left(\frac{2\sqrt{2/3r}}{r_o}\right) \tag{3.4}$$

where M (=3 for fcc polycrystals) is Taylor factor, G_m is the matrix shear modulus, b is the Burgers vector, ν is Poisson's ratio, and r_o is the dislocation core radius.

Eq. (3.4) of Orowan strengthening assumes that the particles are uniformly distributed. The strengthening effect of the particles will be negated if they suffer from inhomogeneous distribution (clustering). In fact the strength of such composites with clustered particles can be inferior compared to the base alloy [25,89] or similar composites with better particle distribution [52]. Clusters induce porosity and reduce the effective volume fraction of reinforcement that can transfer the load [90]. Liu et al. [25] proposed a model to take into account the effect of clustering of reinforcement (CNTs in their case). According to their proposed modified equation that accounts for the cluster and porosity, the strength of a composite (σ_L) is given by

$$\sigma_L = (\sigma_i + kd^{-1/2})\left[(V_f - V_{\text{cluster}})(s+4)/4 + (1-(V_f - V_{\text{cluster}}))\right]\exp(-\psi\theta)] \tag{3.5}$$

where s is the aspect ratio of reinforcement, θ is the porosity content, and ψ is a constant which is related to the shape and size of the pores [91].

REFERENCES

[1] R.S. Mishra, Z.Y. Ma, Mater. Sci. Eng. R 50 (2005) 1–78.

[2] P. Heurtier, C. Desrayaud, F. Montheillet, Mater. Sci. Forum 396–402 (2002) 1537–1542.

[3] R.S. Mishra, Z.Y. Ma, I. Charit, Mater. Sci. Eng., A 341 (2002) 307–310.

[4] Y. Morisada, H. Fujii, T. Nagaoka, M. Fukusumi, Mater. Sci. Eng. A 419 (2006) 344–348.

[5] Y. Morisada, H. Fujii, T. Nagaoka, M. Fukusumi, Mater. Sci. Eng. A 433 (2006) 50–54.

[6] Y. Morisada, H. Fujii, T. Nagaoka, K. Nogi, M. Fukusumi, Composites Part A 38 (2007) 2097–2101.

[7] C.J. Lee, J.C. Huang, P.J. Hsieh, Scripta Mater. 54 (2006) 1415–1420.

[8] C.J. Lee, J.C. Huang, Mater. Trans. 47 (2006) 2773–2778.

[9] K. Sun, Q.Y. Shi, Y.J. Sun, G.Q. Chen, Mater. Sci. Eng. A 547 (2012) 32–37.

[10] W. Wang, Q.-Y. Shi, P. Liu, H.-K. Li, T. Li, J. Mater. Process. Technol. 209 (2009) 2099–2103.

[11] D.K. Lim, T. Shibayanagi, A.P. Gerlich, Mater. Sci. Eng. A 507 (2009) 194–199.

[12] S. Sahraeinejad, H. Izadi, M. Haghshenas, A.P. Gerlich, Mater. Sci. Eng. A 626 (2015) 505–513.

[13] S.A. Alidokht, A. Abdollah-zadeh, S. Soleymani, H. Assadi, Mater. Des. 32 (2011) 2727–2733.

[14] D. Lu, Y. Jiang, R. Zhou, Wear 305 (2013) 286–290.

[15] A. Dolatkhah, P. Golbabaei, M.K.B. Givi, F. Molaiekiya, Mater. Des. 37 (2012) 458–464.

[16] M. Dixit, J.W. Newkirk, R.S. Mishra, Scripta Mater. 56 (2007) 541–544.

[17] D.R. Ni, J.J. Wang, Z.N. Zhou, Z.Y. Ma, J. Alloys Compd. 586 (2014) 368–374.

[18] Q. Liu, L. Ke, F. Liu, C. Huang, L. Xing, Mater. Des. 45 (2013) 343–348.

[19] Z. Du, M.J. Tan, J.F. Guo, G. Bi, J. Wei, Mater. Sci. Eng. A 667 (2016) 125–131.

[20] L. Zhang, R. Chandrasekar, J.Y. Howe, M.K. West, N.E. Hedin, W.J. Arbegast, et al., App. Mater. Interfaces 1 (2009) 987–991.

[21] A. Sharma, B. Vijendra, K. Ito, K. Kohama, M. Ramji, B.V.H. Sai, J. Manuf. Process. 26 (2017) 122–130.

[22] I.S. Lee, C.J. Hsu, C.F. Chen, N.J. Ho, P.W. Kao, Compos. Sci. Technol. 71 (2011) 693–698.

[23] Z.Y. Liu, B.L. Xiao, W.G. Wang, Z.Y. Ma, Carbon 50 (2012) 1843–1852.

[24] Z.Y. Liu, B.L. Xiao, W.G. Wang, Z.Y. Ma, Carbon 62 (2013) 35–42.

[25] Z.Y. Liu, B.L. Xiao, W.G. Wang, Z.Y. Ma, Carbon 69 (2014) 264–274.

[26] J. Qian, J. Li, J. Xiong, F. Zhang, X. Lin, Mater. Sci. Eng. A 550 (2012) 279–285.

[27] L. Ke, C. Huang, L. Xing, K. Huang, J. Alloys Compd. 503 (2010) 494–499.

[28] A. Kumar, R. Raj, S.V. Kailas, Mater. Des. 85 (2015) 626–634.

[29] A. Kumar, D. Yadav, C.S. Perugu, S.V. Kailas, Mater. Des. 113 (2017) 99–108.

[30] Q. Zhang, B.L. Xiao, Q.Z. Wang, Z.Y. Ma, Mater. Lett. 65 (2011) 2070–2072.

[31] Q. Zhang, B.L. Xiao, W.Z. Wang, Z.Y. Ma, Acta Mater. 60 (2012) 7090–7103.

[32] C.J. Hsu, C.Y. Chang, P.W. Kao, N.J. Ho, C.P. Chang, Acta Mater. 54 (2006) 5241–5249.

[33] G.L. You, N.J. Ho, P.W. Kao, Mater. Charact. 80 (2013) 1–8.

[34] G.L. You, N.J. Ho, P.W. Kao, Mater. Lett. 90 (2013) 26–29.

[35] C.J. Hsu, P.W. Kao, N.J. Ho, Scripta Mater. 53 (2005) 341–345.

[36] C.J. Hsu, P.W. Kao, N.J. Ho, Mater. Lett. 61 (2007) 1315–1318.

[37] W.J. Arbegast, P.J. Hartley, in: Proceedings of the Fifth International Conference on Trends in Welding Research, Pine Mountain, GA, USA, June 1–5, 1998, p. 541.

[38] G.J. Buffa, R. Hua, R. Shivpuri, L. Fratini, Mater. Sci. Eng. A 419 (2006) 389–396.

[39] C.I. Chang, C.J. Lee, J.C. Huang, Scripta Mater 51 (2004) 509–514.

[40] M. Bahrami, K. Dehghani, M.K.B. Givi, Mater. Des. 53 (2014) 217–225.

[41] M. Azizieh, A.H. Kokabi, P. Abachi, Mater. Des. 32 (2011) 2034–2041.

[42] A. Devaraju, A. Kumar, B. Kotiveerachari, Mater. Des. 45 (2013) 576–585.

[43] G. Faraji, P. Asadi, Mater. Sci. Eng. A 528 (2011) 2431–2440.

[44] J.F. Guo, J. Liu, C.N. Sun, S. Maleksaeedi, G. Bi, M.J. Tan, et al., Mater. Sci. Eng. A 602 (2014) 143–149.

[45] M. Dadashpour, A. Mostafapour, R. Yeşildal, S. Rouhi, Mater. Sci. Eng. A 655 (2016) 379–387.

[46] S. Hirasawa, H. Badarinarayan, K. Okamoto, T. Tomimura, T. Kawanami, J. Mater. Process. Technol. 210 (2010) 1455–1463.

[47] A. Scialpi, L.A.C. De Filippis, P. Cavaliere, Mater. Des. 28 (2007) 1124–1129.

[48] W.M. Thomas, E.D. Nicholas, S.D. Smith, in: Proc. Aluminum Joining Symposium, TMS Annual Meeting, 11–15 February 2001, New Orleans, Louisiana, USA.

[49] W.M. Thomas, K.I. Johnson, C.S. Wiesner, Adv. Eng. Mater. 5 (2003) 485–490.

[50] R. Rai, A. De, H.K.D.H. Badheshia, T. DebRoy, Sci. Technol. Weld. Join. 16 (2011) 325–342.

[51] C.V. Rao, G.M. Reddy, K.S. Rao, Defence Technol 3 (2015) 197–208.

[52] M. Bahrami, M.K.B. Givi, K. Dehghani, N. Parvin, Mater. Des. 53 (2014) 519–527.

[53] K. Elangovan, V. Balasubramanian, Mater. Sci. Eng. A 459 (2007) 7–18.

[54] V.V. Bhanu Prasad, B.V.R. Bhat, Y.R. Mahajan, P. Ramakrishnan, Mater. Sci. Eng. A 337 (2002) 179–186.

[55] J.E. Spowart, B. Maruyama, D.B. Miracle, Mater. Sci. Eng. A 307 (2001) 51–66.

[56] D.J. Llyod, Acta Metall. Mater. 39 (1991) 59–71.

[57] A.M. Murphy, S.J. Howard, T.W. Clyne, Mater. Sci. Technol. 14 (1998) 959–968.

[58] S.K. Koh, S.J. Oh, C. Li, F. Ellyin, Int. J. Fatigue 21 (1999) 1019–1032.

[59] M. Jeni, M. Kukuchi, Acta Mater. 46 (1998) 3125–3133.

[60] J.L. Lepinoux, Y. Estrin, Acta Mater. 48 (2000) 4337–4347.

[61] J.J. Lewandowski, C. Liu, Mater. Sci. Eng. A 107 (1989) 241–255.

[62] R.J. Arsenault, S. Fishman, M. Taya, Prog. Mater. Sci. 38 (1994) 1–157.

[63] P.B. Berbon, W.H. Bingel, R.S. Mishra, Scripta Mater. 44 (2001) 61–66.

[64] D. Yadav, R. Bauri, J. Mater. Eng. Perform. 24 (2015) 1116–1124.

[65] R. Bauri, D. Yadav, G. Suhas, Mater. Sci. Eng. A 528 (2011) 4732–4739.

[66] P. Kurtyka, N. Rylko, T. Tokarski, A. Wójcicka, A. Pietras, Compos. Struct. 133 (2015) 959–967.

[67] R. Yang, Z. Zhang, Y. Zhao, G. Chen, Y. Guo, M. Liu, et al., Mater. Charact. 106 (2015) 62–69.

[68] W. Hoziefa, S. Toschi, M.M.Z. Ahmed, Al Morri, A.A. Mahdy, M.M. El-Sayed Seleman, et al., Mater. Des. 106 (2016) 273–284.

[69] H.B. Michael Rajan, I. Dinaharan, S. Ramabalan, E.T. Akinlabi, J. Alloys Compd. 657 (2016) 250–260.

[70] Z. Zhang, R. Yang, Y. Guo, G. Chen, Y. Lei, Y. Cheng, et al., Mater. Sci. Eng. A 689 (2017) 411–418.

[71] R. Bauri, Bull. Mater. Sci. 37 (2014) 571–578.

[72] L.M. Marzoli, A.V. Strombeck, J.F. Dos Santos, C. Gambaro, L.M. Volpone, Compos. Sci. Technol. 66 (2006) 363–371.

[73] A. Tewari, J.E. Spowart, A.M. Gokhale, R.S. Mishra, D.B. Miracle, Mater. Sci. Eng. A 428 (2006) 80–90.

[74] H. Izadi, A. Nolting, C. Munro, D.P. Bishop, K.P. Plucknett, A.P. Gerlich, J. Mater. Process. Technol. 213 (2013) 1900–1907.

[75] J.Q. Su, W.T. Nelson, C.J. Sterling, Mater. Sci. Eng. A 405 (2005) 277–286.

[76] K.V. Jata, S.L. Semiatin, Scripta Mater 43 (2004) 743–749.

[77] D. Yadav, R. Bauri, Mater. Sci. Technol. 27 (2011) 1163–1169.

[78] S. Gourdet, F. Montheillet, Acta Mater. 51 (2003) 2685–2699.

[79] F.J. Humphreys, M. Hatherly, Recrystallization and Related Annealing Phenomena, first ed., Pergamon Press, Oxford, 1995.

[80] P.B. Prangnell, C.P. Heason, Acta Mater. 53 (2005) 3179–3192.

[81] J.F. Guo, B.Y. Lee, Z. Du, G. Bi, M.J. Tan, J. Wei, JOM 68 (2016) 2268–2273.

[82] F. Khodabakhshi, A. Simchi, A.H. Kokabi, M. Nosko, F. Simančik, P. Švec, Mater. Sci. Eng. A 605 (2014) 108–118.

[83] I. Charit, R.S. Mishra, Scripta Mater 58 (2008) 367–371.

[84] D. Yadav, PhD Thesis, Indian Institute Technology Madras, India, 2015.

[85] D.J. Lloyd, Int. Mater. Rev. 39 (1994) 1–23.

[86] C.Y. Yu, P.W. Kao, C.P. Chang, Acta Mater. 53 (2005) 4019–4028.

[87] Y.S. Sato, M. Urata, H. Kokawa, K. Ikeda, Mater. Sci. Eng. A 354 (2003) 298–305.

[88] J.W. Martin, Micromechanisms in particle hardened alloys, Cambridge University Press, Cambridge, 1980, p. 60.

[89] M. Barmouz, M.K.B. Givi, J. Seyfi, Mater. Charact. 62 (2011) 108–117.

[90] Z.Y. Liu, Q.Z. Wang, B.L. Xiao, Z.Y. Ma, Compos. Part A: App. Sci. Manuf. 41 (2010) 1686–1692.

[91] S.H. Hong, K.H. Chung, Mater. Sci. Eng. A 194 (1995) 165–170.

CHAPTER 4

Processing Nonequilibrium Composite (NMMC) by FSP

ABSTRACT

Metal matrix composites (MMCs) combine the properties of two different materials to produce a material with superior mechanical properties. Table 4.1 shows the mechanical properties of some of the Al based MMCs. One of the major shortcomings of MMCs, as can be seen in Table 4.1 as well, is the low ductility which arises out of various reasons related primarily to the hard and brittle reinforcement [1–5]. If the primary reason for this be the brittle ceramic particles, can there be alternative reinforcements that can prevent this embrittlement? Harder metallic particles may be the answer. However, the equilibrium phase diagrams say that the metallic particles will either dissolve to form solid solution or react with aluminum to form intermetallic compounds if they have low solubility. Metals can be classified into two groups based on their solid solubility in aluminum as shown in Table 4.2 [6]. Metals from the low solubility group have to be chosen as reinforcement particles as they do not dissolve in aluminum. Ni, Ti, and W are attractive choices as reinforcements because of their higher strength and stiffness compared to aluminum. However, owing to their low solubility such metals will react and form brittle intermetallics when processed in equilibrium conditions by conventional routes. Moreover, the reaction is exothermic in nature that leads to rapid reaction kinetics [7]. Therefore the conventional composite processing routes such as powder metallurgy (PM) and stir casting cannot be used to incorporate metallic particles in aluminum [8–10].

Friction stir processing (FSP) is a solid state processing route that can offer the solution to incorporate metallic particles in their elemental form in aluminum. This chapter presents a detailed overeview of how metal particles are incorporated as reinforcement in an aluminum matix by FSP to process composites. Such composites are defined as "non-equilibrium composites" (NMMCs) since, the metallic particles cannot be retained in their elemental state in an aluminum matrix under equilibrium conditions. Two types of particle reinforcements, one having low solid solubility in aluminum (nickel and titanium) and other having high solid solubility (copper) are used as reinforcement in pure

Metal Matrix Composites by Friction Stir Processing. DOI: http://dx.doi.org/10.1016/B978-0-12-813729-1.00004-8

Table 4.1 Mechanical Properties of Aluminum Based Ceramic Particle Reinforced Composites. Properties of Unreinforced Matrix are Indicated in Brackets

Composite	Yield Strength (MPa)	UTS (MPa)	% of Elongation	Reference
Al356-15 vol.% SiC	324 (228)	331 (287)	0.3 (13)	[11]
Al356-5.5 wt.% TiB_2	240 (228)	303 (287)	7 (13)	[11]
Al2014-SiC	210 (153)	406 (402)	11.5 (21.7)	[12]
Al-3.5Cu-Al_2O_3	238 (174)	374 (261)	2.2 (14.0)	[12]
Al-Li-10 vol.% SiC	372	462	2.0	[13]
A356-4.7 vol.% TiB_2	213 (200)	252 (232)	7.4 (11.1)	[14]
Al-8 vol.% SiC PM	60 (57)	96 (84)	12.5 (26.2)	[15]
Al2024-5 vol.% SiC	214 (75)	320 (185)	3.2 (21)	[16]
Al6061-30 vol.% Al_2O_3	400	533	1.7	[17]
Al7015-5 wt.% B_4C	–	385 (340)	–	[18]
Al356-5 vol.% nano-Al_2O_3	110 (93)	160 (115)	1.7 (2.9)	[19]
Al-Si-6 wt.% ZrO_2	132 (120)	248 (170)	5.8 (11)	[20]
Al-12.3 vol.% Al_2O_3	136 (84)	116 (56)	5.7 (19.3)	[21]

Table 4.2 Maximum Solid Solubility of Various Metals in Aluminum [6]

Low Solid Solubility		
Metal	**Solubility (wt.%)**	**Temperature (°C)**
Au	0.36	640
Co	0.02	660
Cr	0.77	660
Fe	0.05	655
Mn	1.82	660
Mo	0.25	660
Nb	0.22	660
Ni	0.05	640
Sc	0.38	660
Ti	1.0	665
V	0.6	665
Zr	0.28	660
W	0.16	660
High Solubility		
Metal	**Solubility (wt.%)**	**Temperature (°C)**
Ag	55.6	570
Cu	5.67	550
Ga	20	30
Li	4.0	600
Mg	14.9	450
Zn	82.8	380

aluminum matrix. The feasibility of processing nonequilibrium composite in a wrought aluminum alloy (AA5083) with tungsten (W) as reinforcement particles, is also shown. The effect of the metal particles on the microstructure evolution and the mechanical properties is discussed in detail. Finally, the thermal stability of the microstructure of the processed composites against abnormal grain growth (AGG) is evaluated and the role of the metal particles and their own thermal stability is discussed.

4.1 METHODOLOGY

The groove filling method was followed to incorporate metallic particles into the aluminum matrix. The commercially pure (99.5%) aluminum plate that was used as one of the matrix materials had Fe, Si, Zn, and Ti as the major impurities. The other chosen matrix material was a wrought aluminum alloy AA5083 having a nominal composition of 4.2% Mg, 0.6% Mn, 0.2% Si, and 0.2% Fe. A groove of width 1 mm, depth 2 mm, and length 50 mm was cut precisely on the aluminum plates. The groove was then filled with the metal particles and FSP was carried out along the groove with a tool made of M2 steel. Unlike many other studies, no blunt tool was used in this case to close the groove and the composite was made in one step with a single pass of FSP. The morphological characteristics of nickel, titanium, copper, and tungsten powder used as reinforcement and the optimized process parameters adopted to incorporate them into aluminum to process the composites are given in Table 4.3. The process parameters were optimized by carrying out several trials and a lowest possible ratio (ω/v) of tool rotation speed to traverse speed that was just enough to get a defect free stir zone (SZ) was then adopted. The selection of a lowest possible ω/v ratio was to ensure that the heat input is not very high to initiate a reaction between the reinforcement particles and aluminum.

Table 4.3 Particle Morphology and Optimized Process Parameters Adopted to Process Composites

Composite	Al–Ni	Al–Ti	Al–Cu	AA5083-W
Particle morphology	Spherical	Irregular	Spherical	Spherical/globular
Avg. particle size (μm)	70	20	40	10
Tool rotation speed (rpm)	1000	1000	1000	1200
Tool traverse speed (mm/min)	60	30	60	24

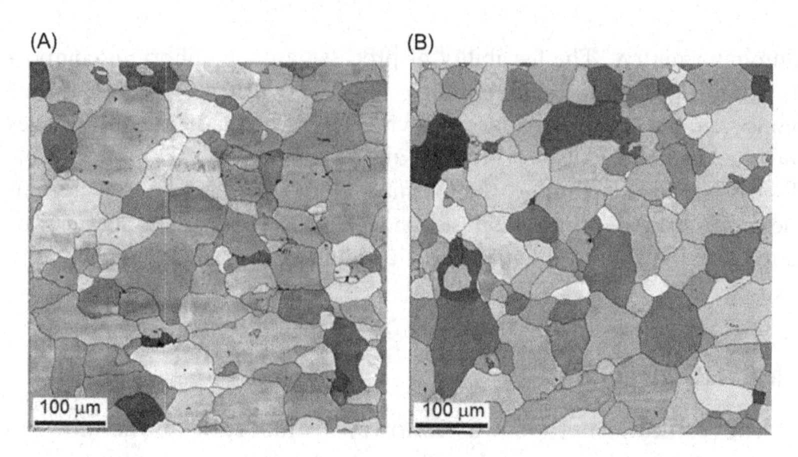

Figure 4.1 EBSD (IPF + grain boundary) map of as-received (A) base Al and (B) 5083 Al plates.

For every composite, FSP was also carried out on the base aluminum plates (without any groove) with the same process parameters for comparison.

The electron backscatter diffraction (EBSD) (IPF + grain boundary) images in Fig. 4.1A and B show the grain size and morphology of the as-received matrix materials. The as-received plates exhibit a coarse equiaxed grain structure. The SEM images of the reinforcement powders (nickel, titanium, copper, and tungsten) are shown in Fig. 4.2A–D. All the particles except Ti had a spherical or globular morphology (Table 4.3).

4.2 Al–Ni COMPOSITE

Among the metals with low solid solubility, nickel is an attractive choice as reinforcement because of its high strength, stiffness, and good high temperature properties. The Al–Ni phase diagram is shown in Fig. 4.3. The maximum solid solubility of nickel in aluminum is 0.05 wt.% at 640°C and it forms a variety of Al–Ni intermetallic such as AlNi, Al_3Ni, Al_3Ni_2, Al_3Ni_5, and $AlNi_3$. Therefore retaining the nickel particles in their elemental state in aluminum is always a challenge. Wong et al. reported processing nickel particle reinforced aluminum composite by the disintegrated metal deposition (DMD) method that vows to reduce particle–liquid contact time to prevent any reaction [22]. However, formation of Al–Ni intermetallics and the consequent reduction in the ductility of the composite could not be avoided.

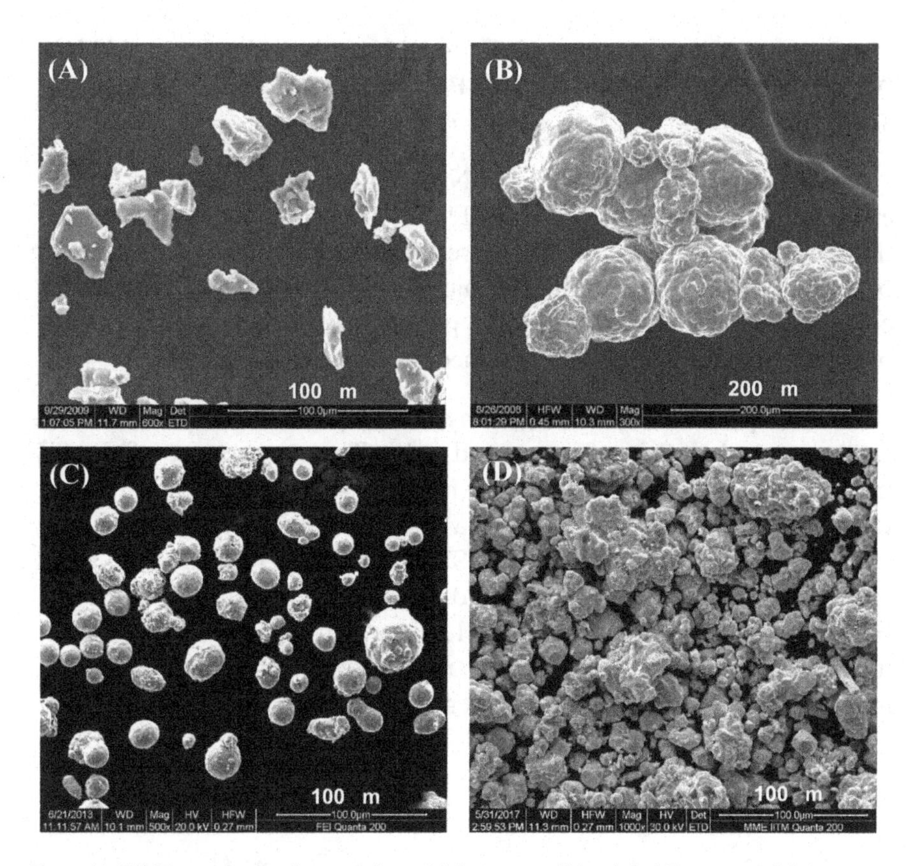

Figure 4.2 SEM image showing the morphology of (A) titanium, (B) nickel, (C) copper, and (D) tungsten particles used as reinforcement.

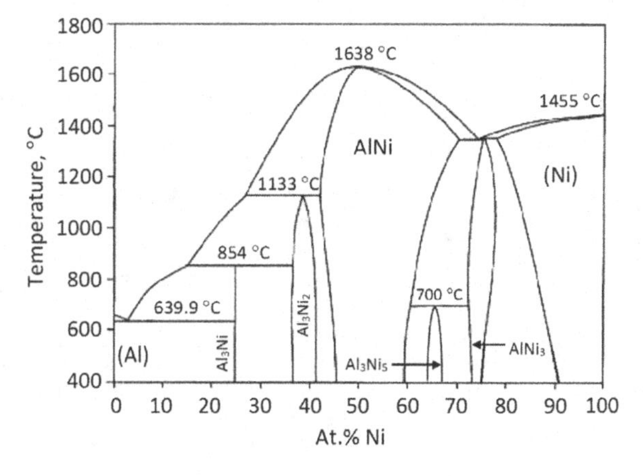

Figure 4.3 Al–Ni binary phase diagram.

PM is another common method for processing aluminum MMCs. Al–Ni intermetallics formed in PM processed Al–Ni composites also [23].

The friction stir processed A–Ni composite was characterized to evaluate the state of the incorporated Ni particles. The X-ray diffraction (XRD) pattern of the processed Al–Ni composite shows peaks belonging to only aluminum and nickel (Fig. 4.4) and no other peaks corresponding to any intermetallic phase was found. Therefore it can be said that Ni particles were retained in their elemental state.

The temperature of the SZ during FSP depends on the ratio of tool rotation speed to the traverse speed (ω/v). The temperature of the SZ in Al has been found to be in the range of 400–500°C [24]. The ω/v ratio in this case, was selected in a such manner that the heat input was not high enough to raise the SZ temperature to a level that can initiate the reaction between Al and Ni. Hence, formation of deleterious intermetallics was prevented. However, a very thin layer of intermetallic, not detectable by XRD, may form at the particle–matrix interface. Nevertheless, it is generally believed that such a layer

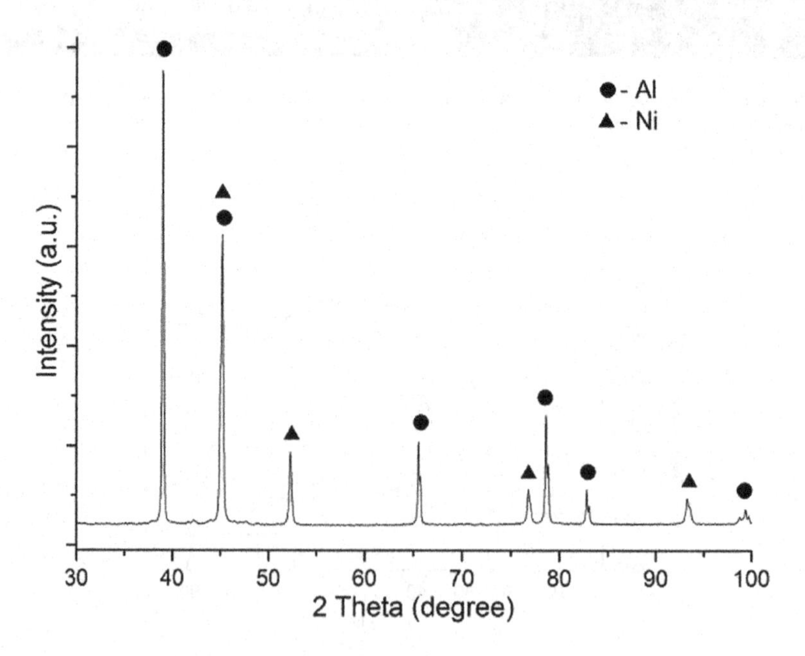

Figure 4.4 XRD pattern of FSPed Al–Ni composite.

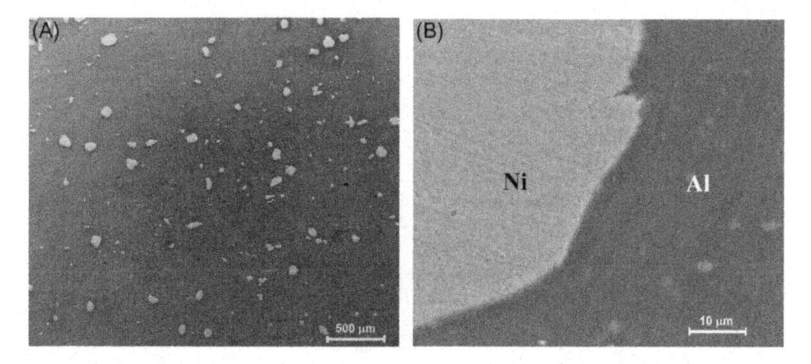

Figure 4.5 SEM micrographs of the Al–Ni composite showing (A) distribution of Ni particles in the stir zone and (B) particle–matrix interface.

improves the bonding between the reinforcement particles and the matrix in MMCs.

The SEM micrographs of the Al–Ni composite are shown in Fig. 4.5. Ni particles were uniformly distributed in the aluminum matrix (Fig. 4.5A). The Ni content was found to be 7 vol.% by image analysis. The particle–matrix interface, as shown in the high magnification image in Fig. 4.5B, was clean with no sign of any particle–matrix reaction and the interfacial bonding was also excellent.

4.3 Al–Ti COMPOSITE

Ti is another attractive choice for reinforcement because of its low density (4.5 g/cm^3) and high specific strength and elastic modulus. The maximum solid solubility of titanium in aluminum is only 1.3 wt.% at 665°C and hence, Ti readily forms Al_3Ti intermetallic with aluminum. Some attempts have been made to prevent the Al–Ti reaction and keep titanium in its elemental state in aluminum by modified liquid metallurgy routes such as disintegrated melt deposition (DMD), where the contact time between aluminum and titanium is minimized [25]. However, formation of intermetallic could not be prevented completely. Even forming an oxide layer on the titanium particles, by prior heat treatment, did not eliminate the intermetallic formation. The result was a decrease in the ductility of the processed composite. The other reported method of incorporating titanium particle into aluminum matrix is accumulative roll bonding (ARB) [26]. However, in this case the samples were very thin as Al foils were used and it

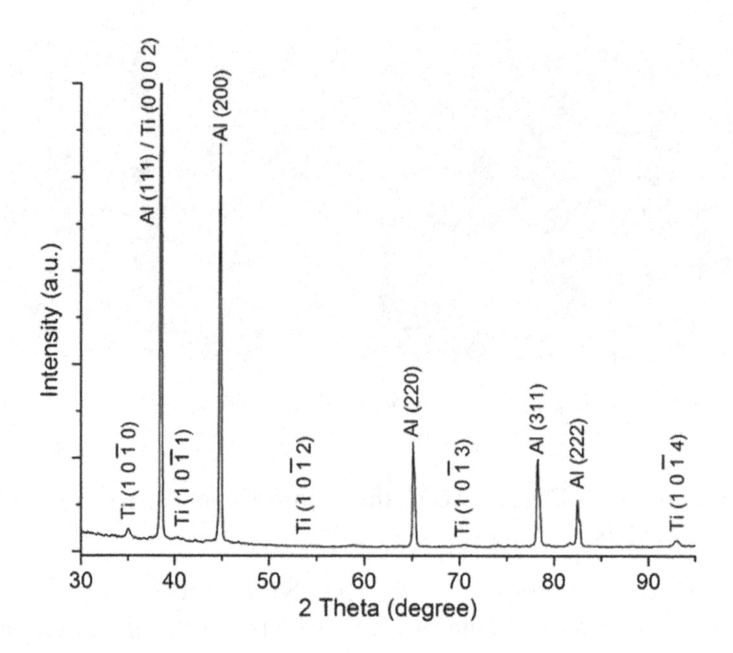

Figure 4.6 XRD pattern of the Al–Ti composite.

required number of ARB cycles (12 cycles) to incorporate the Ti particles. Moreover, the processed composite exhibited low ductility of 5.2%, although the yield strength increased significantly.

FSP could easily incorporate the Ti particle in Al with just a single pass. The XRD pattern of the Al–Ti composite processed by FSP is shown in Fig. 4.6. XRD peaks corresponding to aluminum and titanium alone can be seen. No significant peaks belonging to any intermetallic or any other reaction products were observed confirming that titanium was present in the elemental state.

The surface of the SZ of the composite was analyzed systematically from the advancing side to the retreating side. The SEM micrographs in Fig. 4.7A–C show the microstructure on the advancing side, center, and retreating side, respectively. The particles were uniformly distributed in this case as well. The interface between the composite (SZ) and the base metal was sharp on the advancing side as opposed to a somewhat diffused interface on the retreating side. The particles density on the advancing side was found to be more when compared to the retreating side. This can be attributed to gradients in temperature, strain, and strain rates that exist across the SZ. The material flow

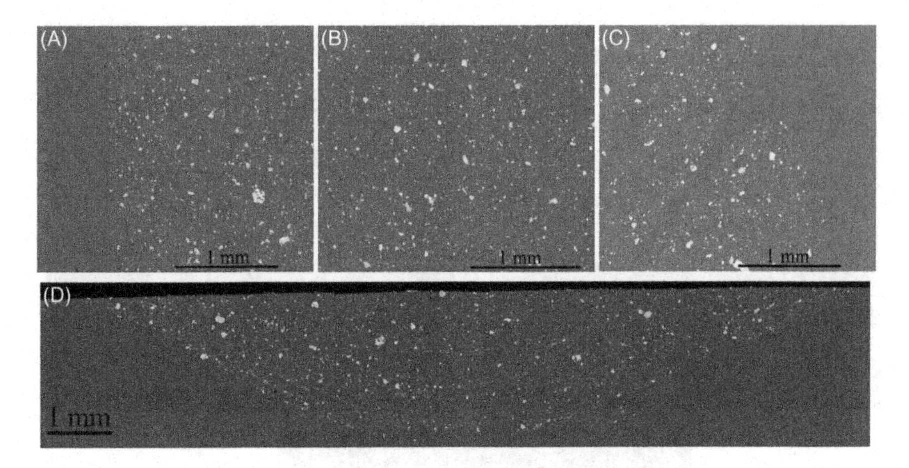

Figure 4.7 SEM (BSE) images of Al–Ti composite on (A) advancing side, (B) center, (C) retreating side. (D) Collage of images showing the cross-section of the stir zone.

during FSP is complex and the flow patterns in advancing and retreating sides are also different giving rise to the gradients in temperature, strain, and strain rate. A collage of the images of the cross-section is shown in Fig. 4.7D. Ti particles delineate the shape of the nugget. It can also be observed that the particles have reached up to the full depth (2 mm) of the groove. The area fraction of Ti in the composite was found to be 7% by image analysis.

The high magnification SEM (BSE) image in Fig. 4.8A shows a sharp particle–matrix interface and no sign of any reaction product forming at the interface. Very few titanium particles were observed with a diffused interface with the matrix on one side as shown in Fig. 4.8B. It appears that there is some dissolution of titanium into the aluminum matrix. However, formation of any equilibrium reaction product does not appear as the contrast difference is gradual. Energy-dispersive X-ray spectroscopy (EDS) phase mapping of the composite shows only two phases, namely elemental Al and elemental Ti and no other phase contrast was observed at the interface as shown in Fig. 4.8C. This substantiates the XRD results and confirms that the particle–matrix reaction is prevented and the titanium particles are retained in their elemental state in the aluminum matrix.

Al_3Ti is the first stable phase which forms after the solubility limit of titanium in aluminum is crossed. A heat input or mechanical energy is needed to overcome the activation barrier for the reaction between

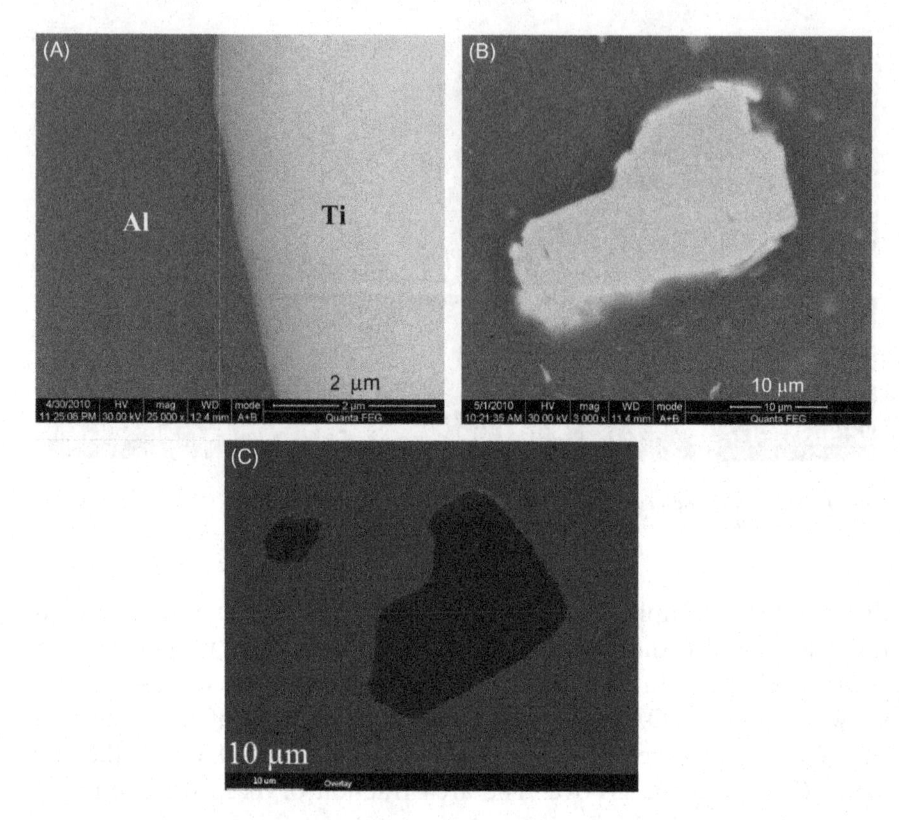

Figure 4.8 SEM (BSE) image showing (A) Al–Ti particle–matrix interface, (B) small Ti particle with diffused interface on one side, and (C) EDS phase mapping.

aluminum and titanium (activation energy = 225 kJ/mol at 823 K [27]). Al_3Ti has a negative heat of formation (−36.9 kJ/atom-mol) [28] and the large heat released is sufficient to cause local melting at the interface and accelerate the reaction. However, this is unlikely to affect if the reaction itself is prevented at the first place as was done in this case by controlling the process parameters of FSP. Using the equation given by Arbegast and Hartley (1999) (Eq. 2.2) temperature in the range of 452–534°C is expected with the parameters used (1000 rpm and 30 mm/minute) in the present case. This temperature corresponds to $0.77–0.86T_m$ for aluminum and $0.37–0.42T_m$ for titanium. Hence, none of the phases go to liquid state during the process. Moreover, the material experiences the peak temperature for a very short duration of time during FSP to have any significant diffusion of atoms and since titanium has very low diffusivity in aluminum [29] the formation of stable Al–Ti reaction products is unlikely. The same argument goes with the processing of Al–Ni composite discussed in Section 4.2. Therefore it can be said that FSP is an effective technique to incorporate insoluble metallic particles as reinforcement without particle–matrix reaction.

4.4 MICROSTRUCTURE AND GRAIN SIZE

The microstructure of the Al–Ni and Al–Ti composites was further characterized by EBSD analysis. The EBSD map of the Al–Ni composite is shown in Fig. 4.9A. A nickel particle can be seen at the center. A fine- and equiaxed-grained microstructure can be observed in the composite. The corresponding grain size distribution is shown in Fig. 4.9B. The average grain size of the matrix of the composite was found to be 7 μm as compared to 74 μm for the unreinforced base metal (Fig. 4.1A). The microstructure development in the Al–Ti composite was similar. However, in order to characterize the microstructure in more detail, EBSD scans were carried out on the advancing side, center, and retreating side and also in the cross-section for the Al–Ti composite. Fig. 4.10A–D show the EBSD (IPF + grain boundary) maps of the composite at different locations. The Al–Ti composite also showed an equiaxed grain structure much finer than the base metal. Fine grains in the range of 4–8 μm were observed in the SZ. The microstructure is also characterized by a high fraction of high-angle grain boundaries (HAGBs) as shown in Table 4.4. The grain size

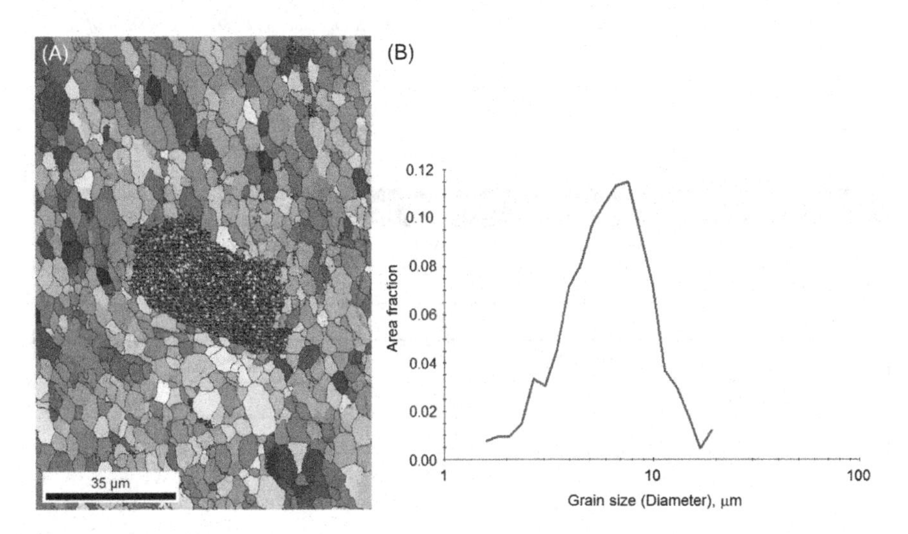

Figure 4.9 (A) EBSD map of the Al–Ni composite and (B) the corresponding grain size distribution.

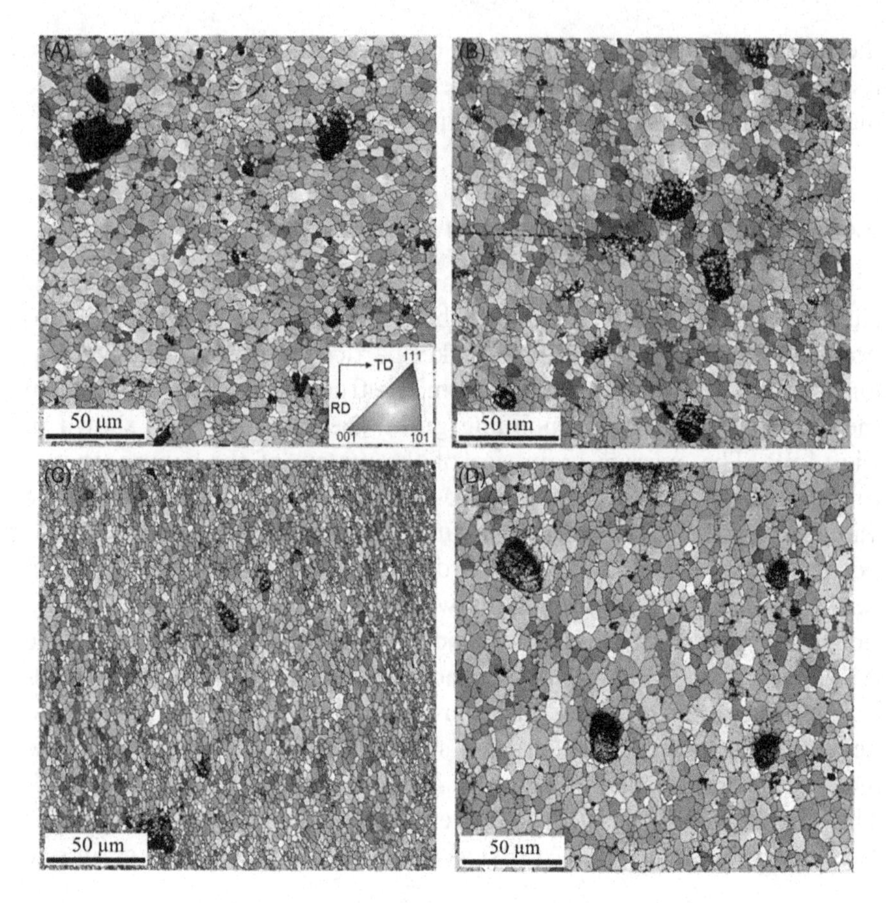

Figure 4.10 EBSD map at (A) advancing side, (B) center, (C) retreating side, and (D) cross-section of Al–Ti composite.

Table 4.4 Summary of Microstructure Developed in Al–Ti Composite and FSPed Al

Al–Ti Composite	**Advancing Side**	**Center**	**Retreating Side**	**Cross-section**
Avg. grain size (μm)	6	8	4	8
% HAGBs	54	42	61	75
FSPed Al	**Advancing Side**	**Center**	**Retreating Side**	**Cross-section**
Avg. grain size (μm)	9	9	7	10
% HAGBs	63	64	72	77

is refined during FSP by dynamic recrystallization. The average grain sizes at different locations are summarized in Table 4.4 and are compared with those of the FSPed Al processed under same conditions. It can be observed that the presence of the particles play a role in refining the grain size due to their pinning effect that restricts the grain growth. Therefore FSP not only incorporated the particles it also refined the grain size significantly.

4.5 Al–Cu COMPOSITE

Processing of nonequilibrium composites by FSP with low solubility metals as reinforcement was discussed in the previous sections. Metals having high solid solubility in aluminum such as zinc, copper, magnesium, and lithium are not suitable as reinforcement and are rather used as alloying elements in aluminum to produce a variety of aluminum alloys such as 2xxx, 7xxx, and 8xxx series. It is demonstrated here that with proper control of the process parameters even a metal from this group can be used as reinforcement to make a nonequilibrium composite by FSP. Cu was chosen for this purpose due to its relatively lower solubility in Al. Higher strength and melting point of Cu compared to aluminum and other soluble metals also make it a suitable reinforcement.

Copper has a maximum solid solubility of 5.67 wt.% (at 550°C) in Al and it forms a stable intermetallic compound, Al_2Cu beyond its solubility limit. There are some reports of making Al/Cu composites in thin strips [30–31] by ARB. The Al–Cu interface was not sharp and the ductility dropped significantly after ARB. According to Liu et al. (2013), aluminum based MMCs with a combination of high strength and good ductility are yet to be produced by ARB [32]. Moreover, ARB involves several cycles and very careful surface preparation between every cycle and the samples produced are very thin. There is no report on processing aluminum copper particulate composite in bulk form. Fabrication of Al–Cu composites by FSP was first demonstrated by Yadav and Bauri [33]. The fabrication approach was the same, i.e., groove filling method.

Cu can go into solution or form Al_2Cu intermetallic depending on the processing temperature and conditions. Hence, it was necessary to closely control and optimize the FSP process parameters to keep Cu as

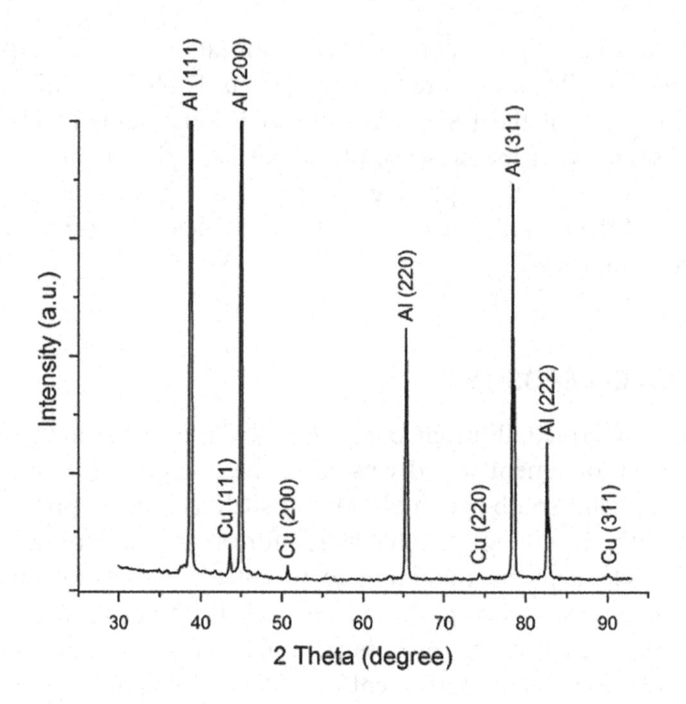

Figure 4.11 XRD pattern of the Al–Cu composite.

elemental particles in Al. The XRD pattern of the processed composite in Fig. 4.11 shows peaks belonging to aluminum and copper and no peak corresponding to any other secondary phase or reaction product was detected.

Fig. 4.12 shows the SEM (BSE) images of the surface and cross-section of the processed composite. A uniform distribution of copper particles in the aluminum matrix can be seen. Area fraction of the particles was found to be around 4%. It can be observed from the cross-section image (Fig. 4.12D) that the particles have gone beyond the groove depth of 2 mm in the center. The particles were observed at high magnification and one such copper particle is shown in Fig. 4.13A. The particle–matrix interface appeared clean without any sign of dissolution of copper in aluminum or formation of any reaction product as shown in Fig. 4.13B. The interfacial bonding is also excellent. It is worth mentioning here that copper has excellent wettability with aluminum and hence, Cu is coated over ceramic particles and carbon fibers to improve their wettability and bonding with the aluminum matrix [34–35]. The EDS elemental overlay of aluminum and copper is shown in Fig. 4.13C. Again uniform distribution of Cu particles can

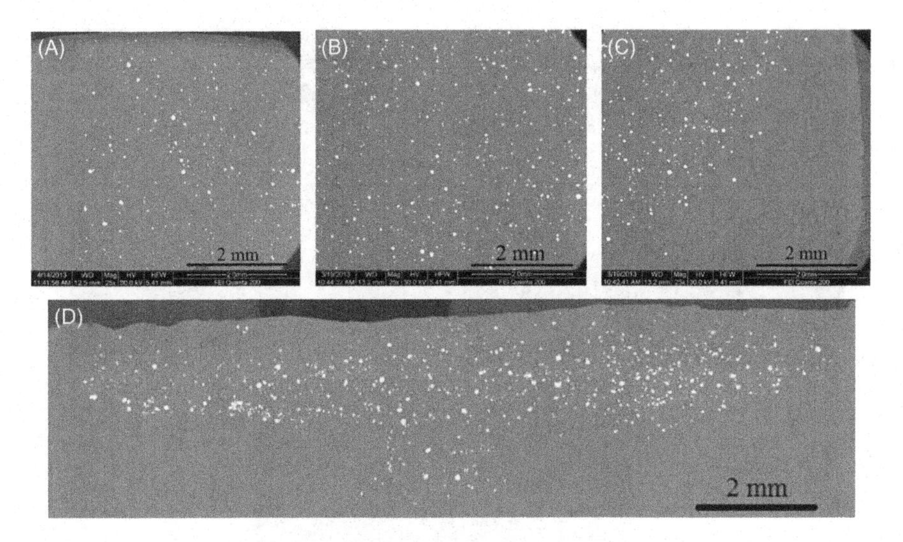

Figure 4.12 SEM (BSE) image of Al–Cu composite on (A) advancing side, (B) center, (C) retreating side, and (D) cross-section.

be seen and no second phase contrast was observed around the particles. The XRD and SEM studies confirm that the copper particles are retained in their elemental state in aluminum.

The grain size of the Al–Cu composite was also analyzed by EBSD. The EBSD (IPF + grain boundary) map of the composite on the surface (center) and in the cross-section is shown in Fig. 4.14A and B, respectively. Equiaxed fine grains along with a copper particle in the center can be seen in the images. The composite exhibited an average grain size of 5 μm with high fraction (>60 %) of HAGBs.

4.6 5083 Al BASED COMPOSITE

In the previous sections, the utility of FSP to process nonequilibrium metal matrix composites (NMMCs) was demonstrated with pure aluminum as the matrix material. Pure aluminum, however, exhibits lower strength and hardness. Therefore the feasibility of using a high-strength alloy as the matrix for preparing NMMCs was considered. Aluminum alloy AA5083 has the highest strength among the wrought alloys and has excellent corrosion resistance and weldability. AA5083, therefore, was chosen as the alloy matrix. In this case, tungsten (W) particles were used as reinforcement since W possesses high stiffness, which is close or even higher than many ceramics, high strength, and good high temperature properties. The FSP parameters however, were

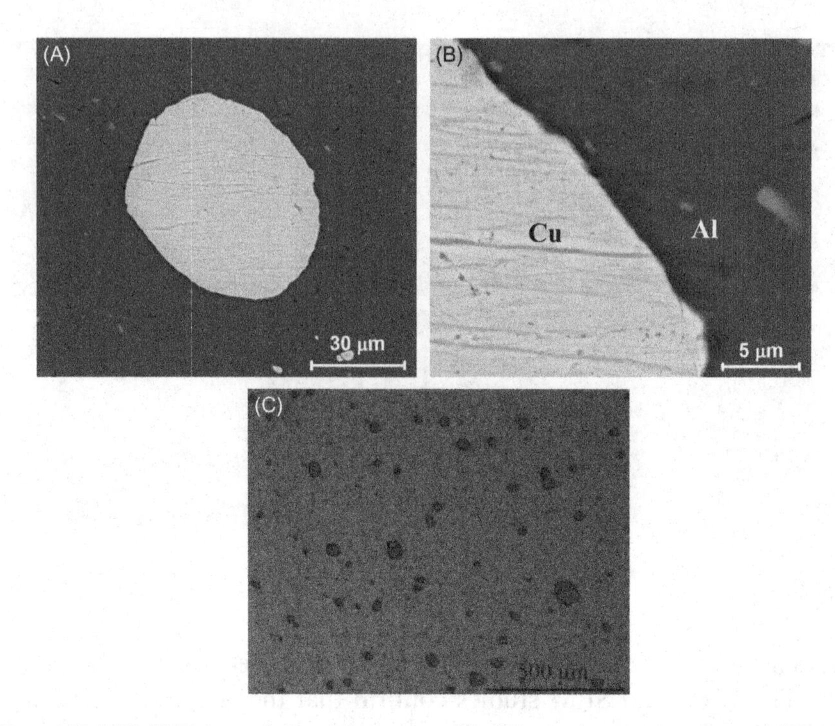

Figure 4.13 SEM (BSE) image showing (A) copper particle, (B) particle–matrix interface, and (C) EDS elemental overlay of the composite.

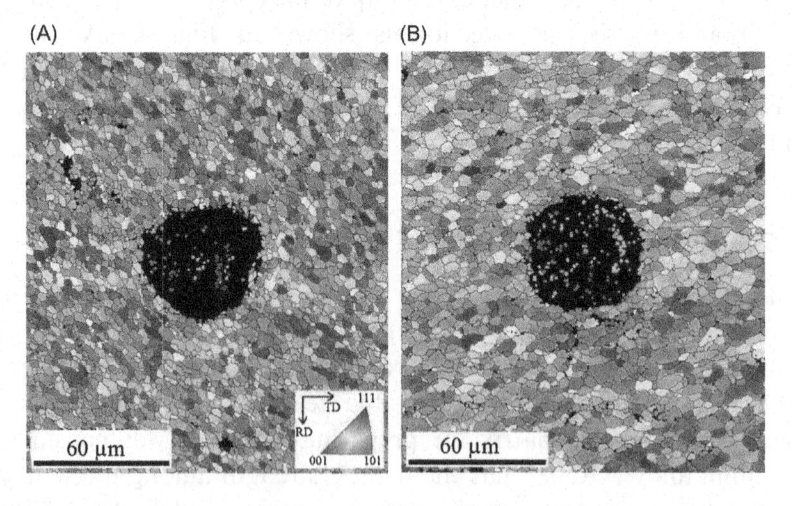

Figure 4.14 EBSD (IPF + grain boundary) map of Al–Cu composite on (A) surface (center) and (B) in the cross-section.

different. Considering the higher strength of the matrix material, a higher rotation speed (1200 rpm) and lower traverse speed (24 mm/minute) compared to the pure Al composites was chosen in order to provide higher heat input. In fact the tool geometry also had to be altered. A plain cylindrical tool was used in case of the pure Al composites. However, a plain tool was not sufficient to cause enough material flow and disperse the W particles in the 5083 Al matrix and hence, a tool with a threaded pin and spiral shoulder profile had to be used. Tungsten has low solid solubility in aluminum and hence, forms a variety of intermetallics, AlW_{12} being the one at low (<10 at.%) tungsten concentrations [36]. Therefore the FSP parameters had to be optimized as well so as not to cause excessive heating that can lead to a reaction between Al and W and formation of Al–W intermetallics.

The XRD pattern of the processed 5083 Al–W composite is shown in Fig. 4.15. Here also peaks belonging to only Al and W can be observed and no other peak corresponding to any intermetallic or any other secondary phase was detected. Therefore the W particles were retained in their elemental state as particulate reinforcement. The SEM (BSE) micrographs in Fig. 4.16A–C show that the W particles are uniformly distributed in the Al matrix on the surface of the SZ. The volume fraction of the particles was found to be around 8%. The cross-section of the processed composite in Fig. 4.16D shows that the W particles

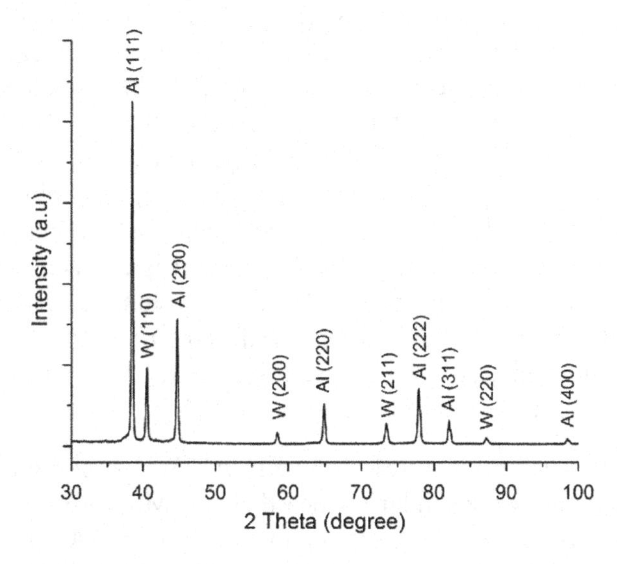

Figure 4.15 XRD pattern of the FSPed 5083 Al–W composite.

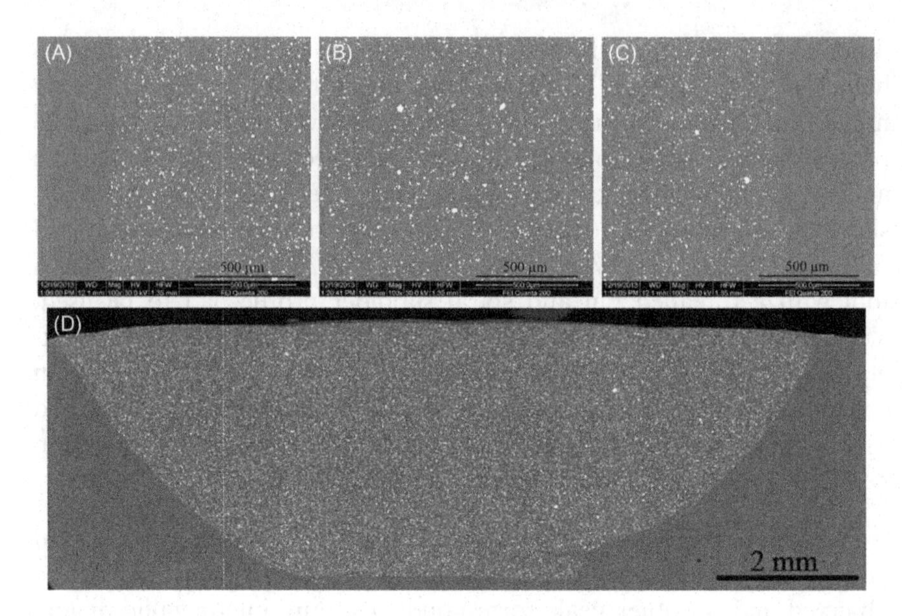

Figure 4.16 SEM (BSE) images showing the distribution of tungsten particles on (A)–(C) the surface, and (D) in the cross-section.

delineate the shape of the nugget and the particles are distributed uniformly along the depth as well. One important aspect that can be noted from the cross-section image is that the W particles have penetrated uniformly to a depth of about 3.5 mm which is far beyond the groove depth (2 mm). This is significant as the particles in composites made by the groove method of FSP are generally limited to the close surface range [37] or maximum to the groove depth [38]. In case of the Al–Cu composite the Cu particles did penetrate more than the groove depth but it was only at the middle part. The deeper penetration of W particles can be attributed to the geometrical profiles (threaded pin and spiral shoulder profile) made on the tool which enhanced the material flow. The volume of material transported by the pin is characterized by the ratio of dynamic volume to static volume, i.e., volume of material swept by the pin to the volume of the pin itself. This ratio is 1.1:1 for plain cylindrical pins while pins with thread-like to triflute kind of geometry can have a ratio in the range of 1.8:1 to 2.6:1 [24].

The microstructural features of the 5083Al–W composite in terms of the grain structure were similar to the other NMMCs processed by FSP. The EBSD map of the composite is shown in Fig. 4.17A and the corresponding image quality map is shown in Fig. 4.17B. A fine-grained microstructure and several tungsten particles (black) can be seen. The

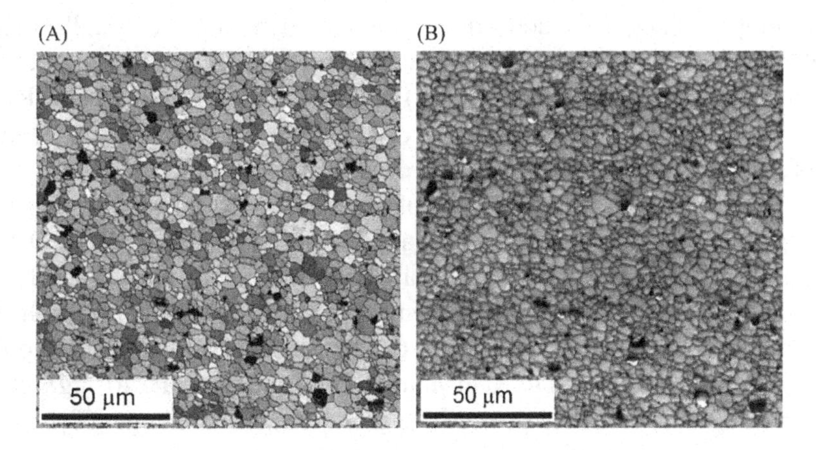

Figure 4.17 (A) EBSD map of the 5083 Al–W composite and (B) corresponding image quality map.

average grain size of the matrix was found to be 4.5 μm as compared to the initial grain size of 25 μm of the base 5083 Al alloy (Fig. 4.1B).

From the microstructure analysis of all the processed NMMCs, it can be seen that FSP has not only incorporated the metal particles in the Al matrix but also caused significant grain refinement of the matrix. The microstructure is characterized by fine and equiaxed grains with a narrow grain size distribution at all the locations of the SZ. The narrow grain size distribution will ensure homogeneous mechanical properties throughout the SZ.

It is worth mentioning here that most of the other SPD processes involve several steps or several passes to achieve a fine grain size. Although grain refinement is achieved, the grain size distribution is generally wide which gives rise to anisotropy in the mechanical properties [39–40]. The grain boundaries are poorly defined and are composed of nonuniformly distributed dislocation structure. Moreover, the crystal lattice is distorted and high internal stresses exist within the grains [41–42].

4.7 MICROSTRUCTURE EVOLUTION IN NMMCs

The microstructure of the composites and FSPed aluminum was further characterized through TEM and EBSD to understand the mechanism of grain refinement during FSP. The grain size is refined by dynamic recrystallization (DRX) process during FSP. Various forms of DRX have been reported as the mechanism of grain refinement during FSP/FSW of

Al alloys. These include discontinuous dynamic recrystallization (DDRX), continuous dynamic recrystallization (CDRX), geometric dynamic recrystallization (GDRX) and dynamic recovery (DRV) [43–48]. DDRX occurs in metals of medium and low stacking fault energy during thermomechanical processes. DDRX takes place by the classical mechanism of nucleation of strain free grains and growth by sweeping motion of grain boundaries [49]. Occurrence of DDRX requires a high density of free dislocations. High rate of recovery in high stacking fault energy materials prevents dislocation accumulation at high strain rates and temperature thereby the possibility of DDRX is reduced. Hence, sustenance of DDRX in aluminum is difficult due to high efficiency of DRV that arises due to its high stacking fault energy [50].

TEM analysis of the NMMCs showed that equiaxed fine grains are bounded by HAGBs as indicated by the extinction contours at the grain boundaries (Fig. 4.18A). The aluminum grains in the composite are found to contain dislocation structures. Some of the grains have high dislocation density as shown in Fig. 4.18B. Subgrain boundaries were also observed inside the grains (Fig. 4.18C). Observations at high magnification revealed that these subgrain boundaries are composed of well-arranged array of dislocations, piled up on a particular plane as

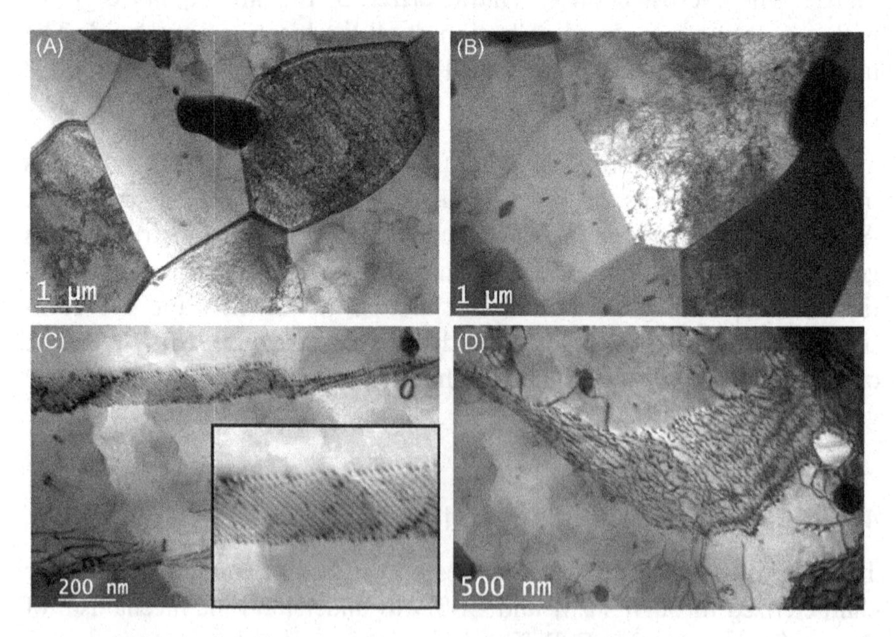

Figure 4.18 TEM images of Al–Ti composite showing (A) fine grains, (B) grains with high dislocation density, (C) subgrain boundaries formed by dislocation array, and (D) dislocation pileup.

shown in the inset. Fig. 4.18D shows dislocation pileup against grain boundaries and variation in the thickness of the subgrain inside a grain. It also shows dislocation absorption into the subgrain boundary (Fig. 4.18D). In the FSPed Al, the dislocation substructures were similar to that of the composite, however, the dislocation content inside the grains was lower than that of the composite. Free dislocations and dislocation tangles were not observed inside most of the grains.

When the grain boundary map of the composites was observed carefully, intermixing of boundaries was observed, i.e., boundaries with mixed orientation were observed. Fig. 4.19 shows the grain boundary map of FSPed Al which exhibits three types of boundaries, i.e., subgrain boundaries (2–5 degrees), low-angle grain boundaries (5–15 degrees) and HAGBs (>15 degrees). Two different contrasts along the same boundary indicates intermixing or mixed character boundaries. White arrows indicate boundaries having partly subgrain character and partly low-angle character and the black arrows indicate intermixing of low-angle and high-angle grain boundaries. The intermixing of boundaries indicates that the misorientation across the boundaries increases gradually that leads to subgrain boundaries converting themselves into low-angle boundaries and the low-angle boundaries eventually getting transformed into HAGBs. The misorientation was individually measured across several such boundaries with mixed character. One such example is shown in Fig. 4.20 where misorientation across three individual boundaries (marked as 1, 2, and 3) was measured.

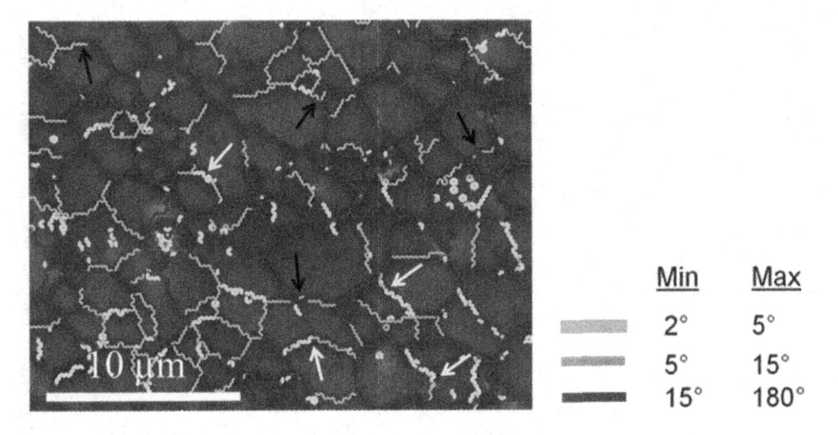

Figure 4.19 Grain boundary map of FSPed Al. White arrows indicate intermixing of subgrain and low-angle boundaries while black arrows indicate low-angle + high-angle boundary mixed character.

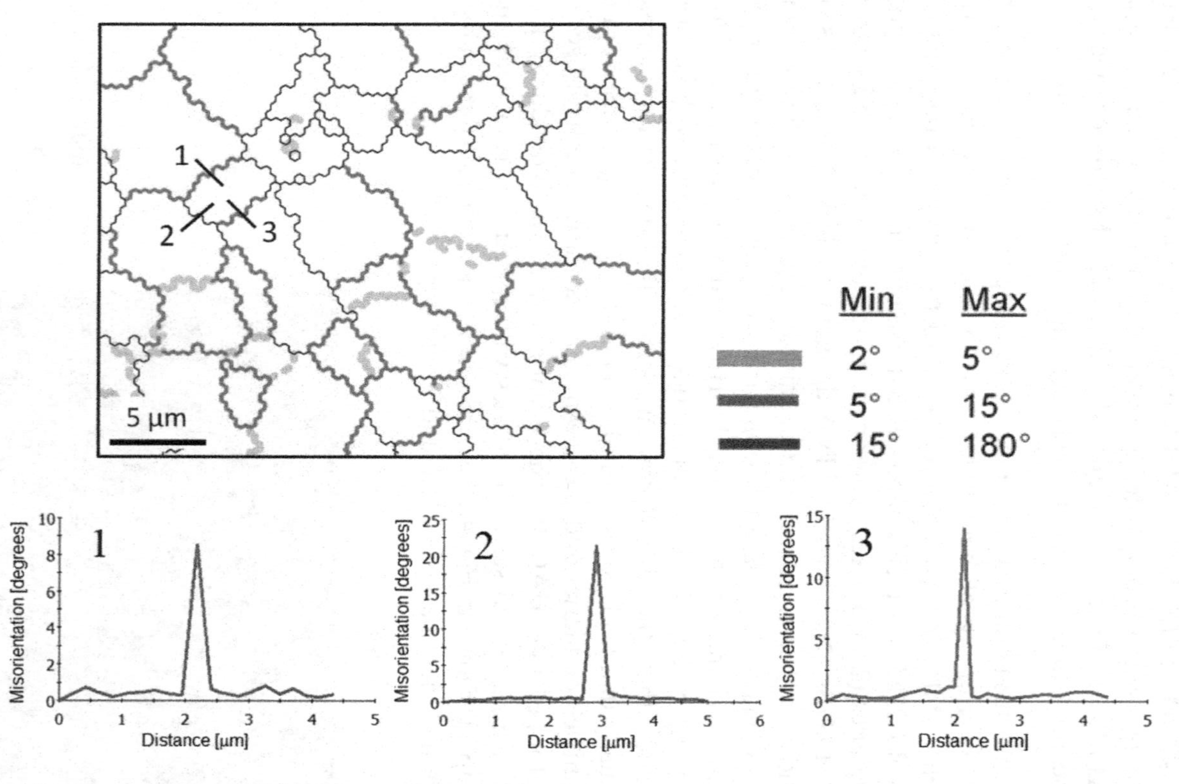

Figure 4.20 Misorientation measured across individual boundaries in the grain boundary map.

The misorientation across boundary "1", which shows single contrast (no intermixing), is found to be 9 degrees and hence, it is confirmed to be a low-angle grain boundary. Similarly the misorientation across boundary "2" is 20 degrees and it complies with the high-angle boundary tag given to it. Boundary "3" is a mixed contrast boundary and shows a misorientation of 14 degrees. The darker contrast in parts of the boundary indicates that the corresponding portions have already become high-angle boundary. Hence, it is not difficult to understand that this boundary ("3") is being converted from low-angle to high-angle boundary.

During FSP as the material is plastically deformed, large numbers of geometrically necessary dislocations (GNDs) are generated. These dislocations rearrange themselves into subgrain boundaries due to high rate of DRV in aluminum. The TEM analysis showed formation of subgrain boundaries and revealed that these boundaries are actually composed of well-arranged array of dislocations (Fig. 4.18C). As the deformation continues in FSP, the dislocations generated get absorbed into the subgrain boundaries, thereby increasing the misorientation across them. Absorption of dislocations into the already formed subgrain boundary is also seen in the TEM images (Fig. 4.18D). Every incoming dislocation that joins the boundary will progressively increase the misorientation across it (some annihilation will also occur) and thus the subgrain boundary gradually transforms into low-angle grain boundary. The low-angle grain boundaries then turn into high-angle boundaries either by subgrain rotation or by further absorption of dislocations. Diffraction contrast is observed across some of the boundaries as shown in Fig. 4.21A and B. This indicates that the misorientation across these subgrain boundaries has increased and is close

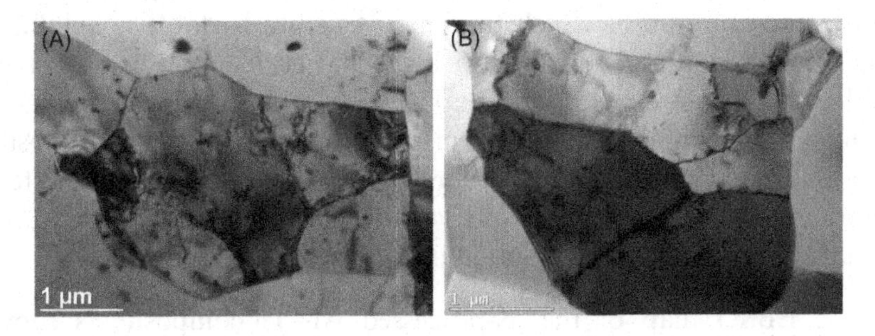

Figure 4.21 (A) and (B) TEM micrograph showing diffraction contrast across subgrain boundaries inside a grain.

to high angle. A DDRX microstructure on the other hand exhibits a recrystallized grain structure free of dislocations and substructures as it takes place by grain boundary migration. Therefore in the present case CDRX driven by DRV seems more likely to operate during FSP. However, it should be noted that the observations are solely based on the final microstructure developed after FSP and several mechanisms or steps may be involved at different stages of the microstructure development. Also, the mechanism of grain refinement may vary from one alloy to other depending on its purity, alloying elements and nature of strengthening precipitates. The microstructural features of FSPed aluminum were similar to that of the composites and it can be said that incorporation of the metallic particles does not affect the mechanism of grain refinement of the aluminum matrix during FSP.

4.8 THERMAL STABILITY OF NMMCs

4.8.1 Thermal Stability of the Microstructure

Fine-grained microstructures are often susceptible to abnormal grain growth (AGG) when subjected to thermal cycles. AGG of FSPed 7075 aluminum was also reported by Charit and Mishra [51]. The driving force for AGG is the reduction in the grain boundary energy. AGG originates by the preferential growth of a few grains which have some special growth advantage over their neighbors. According to recent studies conducted by Omori et al. and Taleff and Pedrazas, the mechanisms that cause AGG are still not well understood [52–53]. Omori et al. suggested that the subgrain boundary energy might be one of the driving forces for AGG [52].

Thermal stability of the microstructure is therefore an important aspect in fine-grained FSPed materials. A prior knowledge of the time and temperature up to which the microstructure is stable is helpful to find the temperature range in which the FSPed material can be safely used or further processed for shaping or manufacturing. In order to evaluate the thermal stability, the processed nonequilibrium composites were subjected to thermal exposure for 10 minutes at different temperatures on the same area followed by EBSD of the exposed area. Results of Al–Ti composite are mainly presented here.

The EBSD map of the as-processed Al–Ti composite is shown in Fig. 4.22A which shows a fine equiaxed grain structure. The

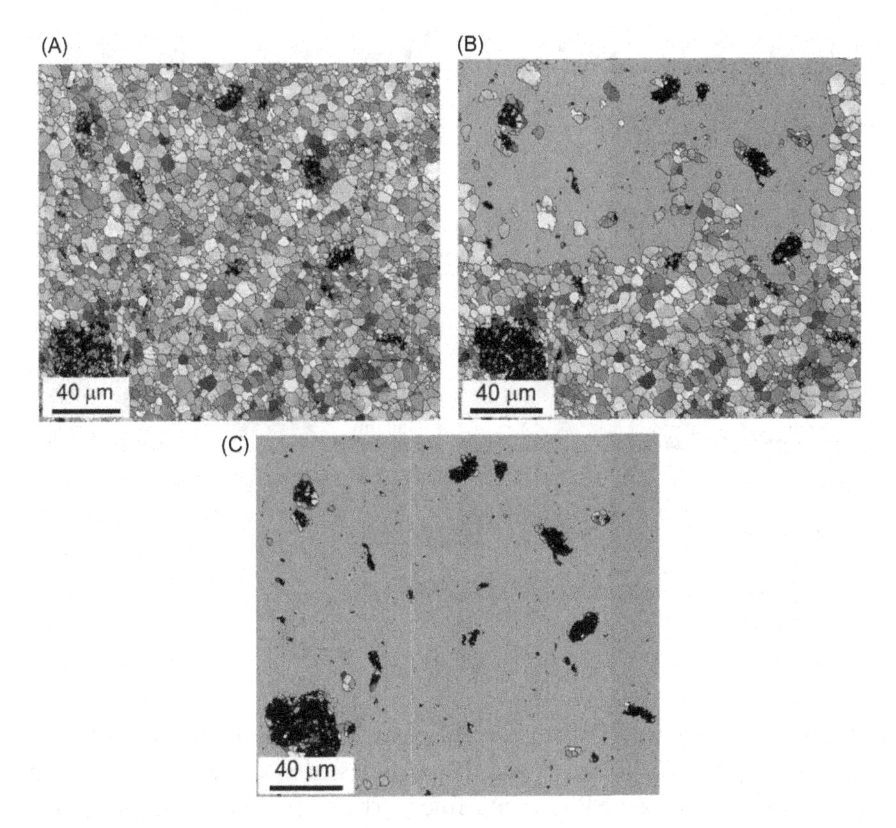

Figure 4.22 EBSD map of Al–Ti composite (A) before heat treatment, (B) after heat treatment at 440°C for 10 min, and (C) 440°C for 20 min.

fine-grained microstructure was stable up to 430°C for 10 minutes exposure. After a thermal exposure at 440°C for 10 minutes AGG started as shown in Fig. 4.22B. During the next 10 minutes (i.e., total 20 minutes of exposure) of exposure the abnormally growing grains increased in size at the expense of the finer grains and eventually consumed the finer grains completely as shown in Fig. 4.22C. There was no sign of any reaction of Ti particle with the aluminum matrix at this temperature and the particle–matrix interface was also clean. The FSPed Al was also given similar thermal treatment and compared with the composite. The EBSD map of FSPed Al before heat treatment is shown in Fig. 4.23A. The FSPed Al underwent AGG at 400°C after 10 minutes of exposure as shown in Fig. 4.23B. Therefore the presence of Ti particles raised the temperature of AGG by 40°C. This can be attributed to the pinning of the grain boundaries by the Ti particles, especially the finer ones. Though the average particle size of the

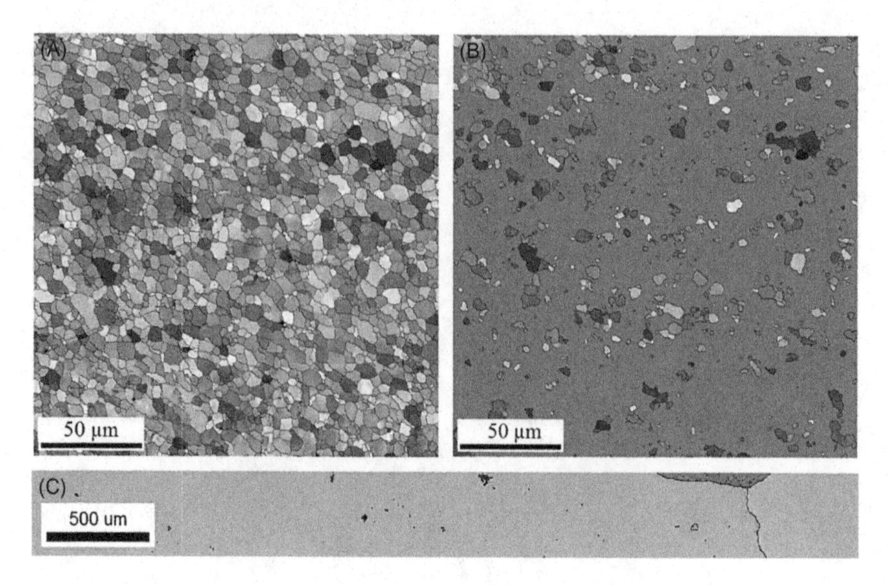

Figure 4.23 EBSD map of FSPed Al (A) before heat treatment, (B) after thermal exposure at 400°C for 10 min, and (C) 400°C for 120 min.

titanium powder was 20 μm, several particles in the range of 5–10 μm were also present. Some of the Ti particles also underwent twinning and fracture during FSP creating fine particles [54]. These small particles can effectively pin the boundaries and delay the AGG or raise the temperature of AGG. Similar improvement in thermal stability of FSPed AZ31 alloy from 300°C to 400°C after incorporating SiC particle was reported by Morisada et al. [55]. Many island grains can be observed in the abnormally grown microstructure of FSPed Al (Fig. 4.23B). One interesting aspect that was noted in the FSPed Al is that on prolonged exposure of 120 minutes at 400°C the grains grew as big as 3–4 mm. This is shown in the large area scan in Fig. 4.23C. By raising the temperature or increasing the exposure time much bigger grains can possibly be produced. Hence, thermal treatment of FSPed microstructure may be suitable for producing large-grained metallic materials in solid state.

In the case of the Al–Cu composite, AGG was observed at 400°C after an exposure of 10 minutes which was same as the FSPed Al. The copper particles were perhaps too big (40 μm) to cause any pinning effect on the grain boundaries and hence could not raise the temperature of AGG when compared with FSPed aluminum. For 5083 Al–W composite, the AGG occurred at 460°C after 10 minutes exposure. Therefore the alloy chemistry and the type of metallic reinforcement seem to play a role in the AGG of NMMCs processed by FSP.

4.8.2 Thermal Stability of the Particles

The metallic particles incorporated in Al matrix are in nonequilibrium state and can react with aluminum or dissolve into aluminum if exposed to thermal cycles, especially during high temperature annealing. Hence, it is important to evaluate the thermal stability of the particles as well. The composites were given a series of heat treatment and the particles were monitored by SEM observations for the particle–matrix reaction.

In the Al–Ti composite, the Ti particles were stable up to a temperature of 530°C. This was much higher than the AGG starting temperature (440°C) in the composite. The particle–matrix reaction started at 550°C. A thermal exposure at 550°C for 20 minutes led to development of a thin layer of reaction product around the titanium particles as shown in Fig. 4.24A. The same particle before heat treatment is shown in the inset. The thickness of the layer varied from 500 nm to almost 1 μm. The composition of the layer was found to be close to that of Al_3Ti intermetallic from EDS analysis. The fully developed and continuous intermetallic layer formed a *core–shell* type structure around the particles as shown for the small particle in Fig. 4.24B. Holding the sample at 550°C for 120 minutes led to the growth of the layer around the particles. Pores were developed in the aluminum matrix as shown in Fig. 4.24C. Pores as big as 10 μm can be found after the heat treatment. Very fine pores were observed in the titanium particles as shown in Fig. 4.24D.

The formation of the intermetallic layer can be attributed to development of a diffusion couple between the aluminum matrix and the metallic particle during the thermal exposure. Melting point of aluminum (663°C) is lower than that of titanium (1668°C). Hence, it is easy to create vacancy in Al than in Ti at high temperature. The Ti atoms diffuse into aluminum through the interface thereby leaving behind vacancies. The vacancies then condense to form pores by concomitant occurrence of Kirkendall effect [56]. The intermetallics have a highly ordered structure and different molar volume capacity than Al and Ti. For example, three Al atoms are needed per titanium atom to form

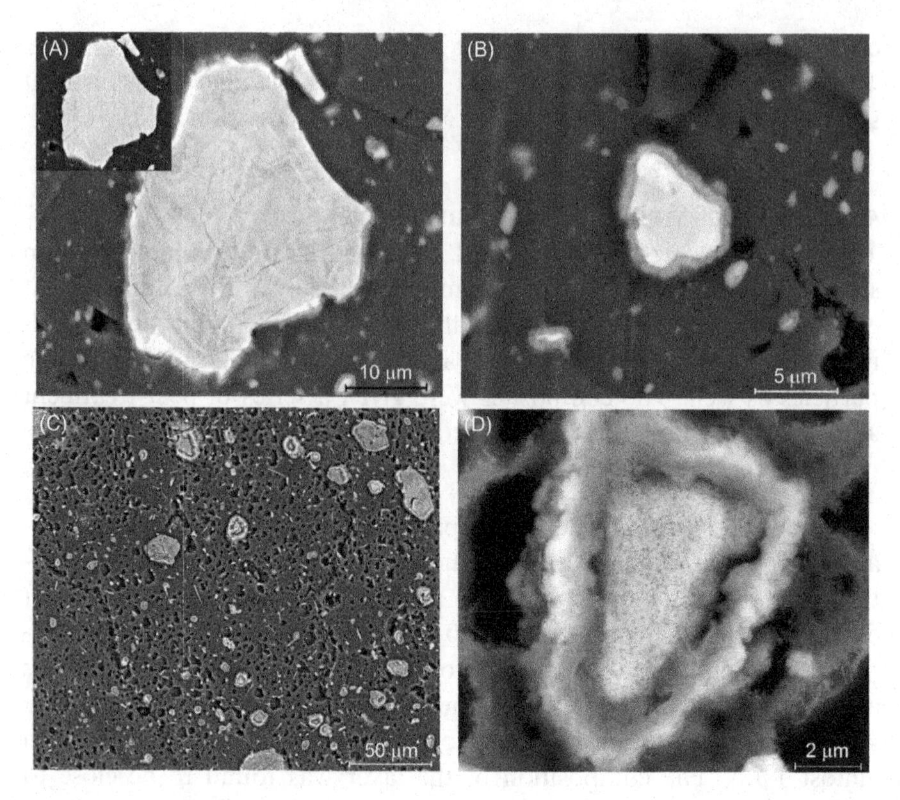

Figure 4.24 SEM (BSE) images of (A) titanium particle after heat treatment at 550°C for 20 min (inset shows the same particle before heat treatment), (B) core–shell *type structure around a small titanium particle, (C) pores in the aluminum matrix after heat treatment at 550°C for 120 min, and (D) fine pores in the titanium particles.*

Al_3Ti kind of intermetallics. Hence, more Al flux towards the interface and consequently more vacancy flux toward aluminum is expected to occur giving rise to larger pores in aluminum.

It may be noted that in the Al–Ti composite the onset temperature (550°C) of particle–matrix reaction is well above the AGG temperature (440°C) and also higher than the solutionizing temperature of most of the Al alloys.

For the Al–Cu composite, the matrix microstructure was more stable than the copper particles. In this case, the particle–matrix reaction occurred at 350°C which is lower than the AGG temperature of the composite (400°C). A complete *core–shell* type structure with a

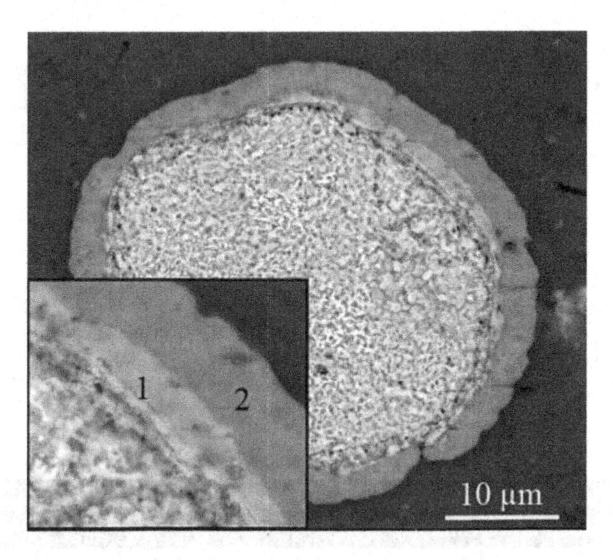

Figure 4.25 SEM (BSE) image of a copper particle after heat treatment at 350°C for 30 min showing the core–shell *type structure that formed around the particle. Inset shows that the* shell *consists of two layers.*

thin continuous layer of around 3 μm formed around the Cu particles at 350°C after 30 minutes exposure (Fig. 4.25). The *shell* consists of two layers (marked 1 and 2) having thickness of around 1 μm and 2 μm, respectively, as shown in Fig. 4.25. EDS analysis revealed that the composition of the outer layer (2) was close to that of Al_2Cu and that of the inner layer (1) was close to AlCu [33]. The center portion still remained as elemental Cu. The formation of a large number of pores in the Cu particles indicates occurrence of Kirkendall diffusion in this case as well. It is worth mentioning here that for the Al–Cu composite, the thickness of the intermetallic layer can be controlled by the time–temperature combination and this may help tailoring the properties without compromising the fine grain structure of the composite since the particle–matrix reaction temperature (350°C) is lower than the AGG (400°C) temperature of the matrix.

4.9 MECHANICAL PROPERTIES

Tensile tests were carried out on the composites as well as FSPed Al (without particles) and base aluminum to evaluate the mechanical properties. The tests were carried out on standard ASTM subsize

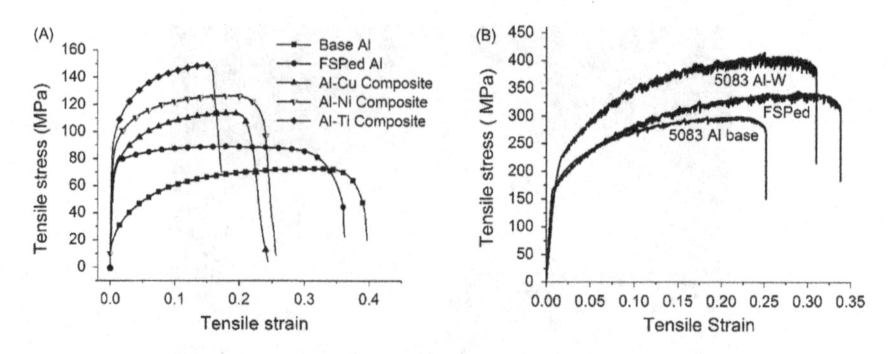

Figure 4.26 Engineering stress–strain curves of (A) pure aluminum based composites and (B) 5083 Al based composite.

Table 4.5 Tensile Properties of Base, FSPed Al, and the Composites (Standard Deviation is Given in Parenthesis)

Material	0.2% Proof Stress (MPa)	UTS (MPa)	% of Elongation	*n*
Base Al	35 (0.6)	72 (1)	39 (0.6)	0.37
FSPed Al	82 (2)	89 (3)	35 (2)	0.15
Al–7% Ti composite	118 (2)	149 (0.7)	16 (0.9)	0.2
Al–7% Ni composite	104 (3.7)	127 (7)	24 (2.6)	0.2
Al–4% Cu composite	88 (1.4)	114 (2.1)	22 (0.7)	0.22
Base AA5083	156 (4)	296 (2)	25 (1)	0.26
FSPed AA5083	169 (6)	337 (5)	33 (1)	0.31
AA5083–8% W composite	214 (10)	404 (23)	30 (1)	0.32

samples (10 mm gage length and 40 mm overall length) which were sliced from the SZ parallel to the surface by electrical discharge machining. A strain rate of 10^{-3}/s was used for the tensile tests. Fig. 4.26A and B shows the engineering stress–strain curves of the pure aluminum based and 5083 Al based composites, respectively. The properties are summarized in Table 4.5.

It can be seen that incorporation of Ti, Ni, and W particles improves the strength compared to both base aluminum and FSPed aluminum. Copper particles improve the strength less significantly as they have the least strength (and hardness) compared to other metal particles. The improvement in strength comes from a combined effect of grain refinement and the particle reinforcement. Orowan strengthening and grain refinement are the main strengthening mechanisms. The metal particles are well-bonded with the matrix and the load is effectively transferred to them through the interface. The grains have high dislocation density due to differential deformation and thermal mismatch between Al and the metal particles. Hence, high dislocation density and subgrain boundaries observed inside the grains also contribute to strength of the composite as they act as hindrance to the dislocation motion.

The important point to note here is that even after a significant improvement in strength the composites retained an appreciable amount of ductility. The absence of any brittle intermetallic phase and excellent interfacial bonding helped retaining the ductility in the composites. The tortuous path to the crack propagation due to the presence of large number of grain boundaries in the fine-grained composites also contribute to the enhanced plastic flow and ductility. The Al–Ti composite is relatively less ductile than Al–Ni and Al–Cu composites. Titanium being an HCP metal has limited number of slip systems and displays much lower ductility than Ni and Cu. As stated before, the Ti particles were also subjected to twinning and fracture during FSP and were strain hardened. These conditions may lead to early fracture of the particles and subsequently failure of the composite during tensile loading. The 5083 Al–W composite showed higher ductility even than the unreinforced alloy. The alloy contained second phase particles which were needle shaped as shown in Fig. 4.27A. These particles are converted into spherical and globular shape after FSP (Fig. 4.27B). The morphology change relaxes the stress concentration points that existed around needle edges and ends and thus

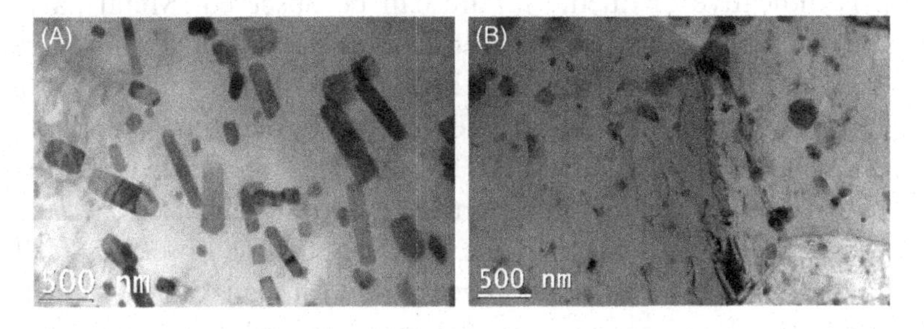

Figure 4.27 TEM micrographs of 5083 Al alloy (A) before and (B) after FSP.

enhances the ductility. The reliving of the residual stresses during FSP may also contribute to the enhanced ductility.

The engineering stress–strain curve was converted into true stress–strain curve to evaluate the strain hardening behavior of the materials. The strain hardening was estimated using the Hollomon equation ($\sigma = K\varepsilon^n$) which gave best fit straight line on a log-log plot in the plastic range. The strain hardening exponent (n) values are reported in Table 4.5. The FSPed 5083 Al and 5083 Al–W composite exhibit higher n values compared to the base alloy. This can be attributed to dislocation structures generated during FSP. A higher density of free dislocations is needed for significant strain hardening to occur. As seen in Fig. 4.27B, the second phase particles present in the alloy pin down many of the dislocations generated during the deformation in FSP and retain them in the recrystallized grains. In case of the composite, the W particles, especially the finer ones, can also pin the dislocations. Moreover, the matrix of the composite contains higher dislocation density due to the thermal mismatch between the reinforcement particles and the Al matrix. In the absence of the fine pinning particles and additional sources of dislocations in pure Al, the strain hardening generally scale with the grain size. Grain boundaries act as both source and sink for dislocations. In a coarser grain material the dislocation annihilation rate is low due to larger distance between the boundaries. This results in higher strain hardening as was observed in the coarse-grained base Al (Table 4.5).

The fracture surface of all the samples was observed under SEM. Fig. 4.28A and B shows the SE and BSE images, respectively, of the fracture surface of the Al–Ti composite. A dimpled fracture surface which is indicative of ductile failure can be observed. Metal particles sitting inside the dimples can be seen from the BSE image (Fig. 4.28B). The high magnification image in Fig. 4.28C shows that there is no decohesion between the reinforcement particles and the Al matrix due to the strong interfacial bonding. To get some more insight into the role of particle–matrix bonding on the fracture behavior, the surface of the broken tensile sample in the gage length was observed in SEM (Fig. 4.28D). The observation reveals that though the Al matrix is deformed, the bonding between the particle and the matrix is still intact and as a result the crack propagates through the particle and not through the particle–matrix interface.

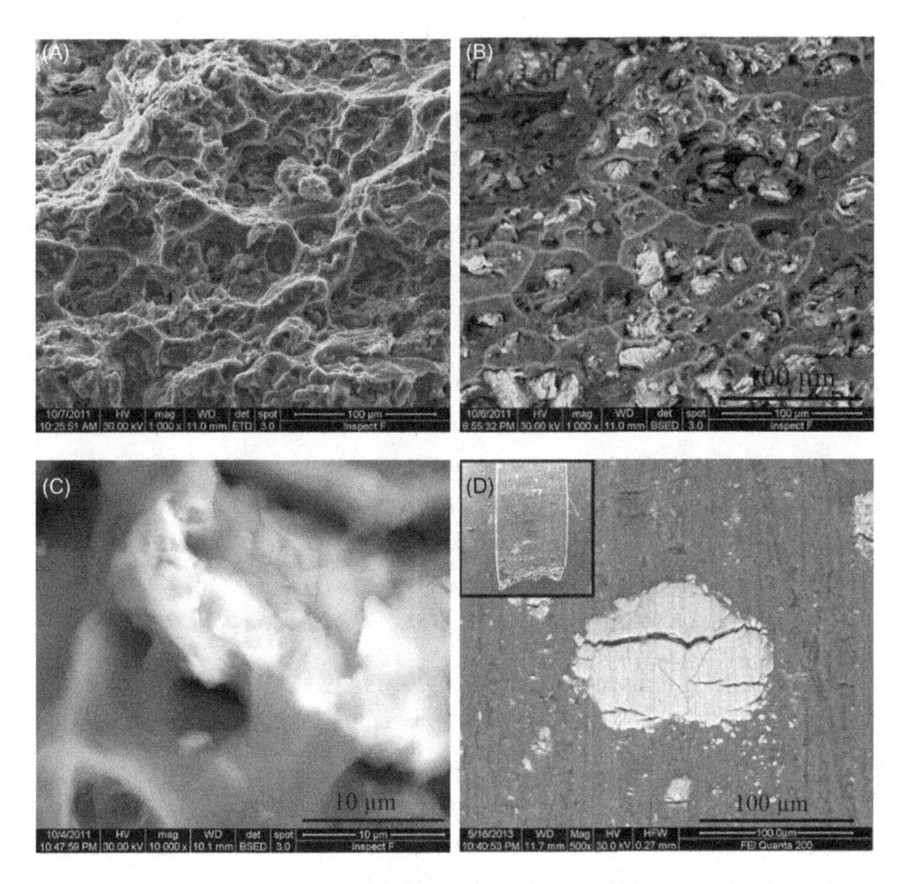

Figure 4.28 SEM images showing the fracture surface of (A) Al–Ti composite, (B) BSE image showing particles sitting in the dimples, (C) particle–matrix bonding, and (D) surface of the broken tensile sample showing cracks propagating in the Al matrix and through the particles.

REFERENCES

[1] T.W. Clyne, P.J. Withers, An introduction to metal matrix composites, first ed., Cambridge University Press, Cambridge, 1993.

[2] R.J. Arsenault, N. Shi, Mater. Sci. Eng. A 81 (1986) 175–187.

[3] J.J. Lewandowski, C. Liu, Mater. Sci. Eng. A 107 (1989) 241–255.

[4] K.B. Lee, H.S. Sim, S.Y. Cho, H. Kwon, Mater. Sci. Eng. A 302 (2001) 227–234.

[5] Z. Wang, M. Song, C. Sun, D. Xiao, Y. He, Mater. Sci. Eng. A 527 (2010) 6537–6542.

[6] T.B. Massalski, H. Okamoto, Binary Alloy Phase Diagrams, ASM International, Materials Park (OH), 1993.

[7] S. Suwas, G.S. Upadhyaya, Met. Mater. Process 7 (1996) 225–250.

[8] L. D'Angelo, L. D'Onofrio, G. Gonzalez, J. Alloys Compd. 483 (2009) 154–158.

[9] M. Krasnowski, T. Kulik, Intermetallics 15 (2007) 201–205.

[10] K. Bouché, F. Barbier, A. Coulet, Mater. Sci. Eng. A 249 (1998) 167–175.

[11] J.V. Wood, P. Davies, J.L.F. Kellie, Mater. Sci. Technol. 9 (1993) 833–840.

[12] B. Ralph, H.C. Yuen, W.B. Lee, J. Mater. Process. Technol. 63 (1997) 339–353.

[13] K. Hanada, Y. Murakoshi, H. Negishi, T. Sano, J. Mater. Process. Technol. 63 (1997) 405–410.

[14] M. Wang, D. Chen, Z. Chen, Y. Wu, F. Wang, N. Ma, Mater. Sci. Eng. A 590 (2014) 246–254.

[15] S. Min, Trans. Nonferrous Met. Soc. China 19 (2009) 1400–1404.

[16] S.J. Hong, H.M. Kim, D. Huh, C. Suryanarayana, B.S. Chun, Mater. Sci. Eng. A 347 (2003) 198–204.

[17] B.F. Luan, N. Hansen, A. Godfrey, G.H. Wu, Q. Liu, Mater. Des. 32 (2011) 3810–3817.

[18] J. Onoro, M.D. Salvador, L.E.G. Cambronero, Mater. Sci. Eng. A 499 (2009) 421–426.

[19] A. Mazahery, H. Abdizadeh, H.R. Baharvandi, Mater. Sci. Eng. A 518 (2009) 61–64.

[20] J. Hemanth, Mater. Sci. Eng. A 507 (2009) 110–113.

[21] J. Singh, S.K. Goel, V.N.S. Mathur, M.L. Kapoor, J. Mater. Sci. 26 (1991) 2750–2758.

[22] W.L.E. Wong, M. Gupta, C.Y.H. Lim, Solid State Phenom. 111 (2006) 39–42.

[23] S.K. Pradhan, S.K. Shee, A. Chanda, P. Bose, M. De, Mater. Chem. Phys. 68 (2011) 166–174.

[24] R.S. Mishra, Z.Y. Ma, Mater. Sci. Eng. R 50 (2005) 1–78.

[25] S.K. Thakur, M. Gupta, Compos. Part A 38 (2007) 1010–1018.

[26] D. Yang, P. Cizek, P. Hodgson, C. Wen, Scripta Mater. 62 (2010) 321–324.

[27] Q. Zhang, B.L. Xiao, Z.Y. Ma, Mater. Chem. Phys. 139 (2013) 596–602.

[28] J.C. Schuster, M. Palm, J. Phase Equilib. Diff. 27 (2006) 255–277.

[29] S.K. Das, L.A. Davis, Mater. Sci. Eng. 98 (1988) 1–12.

[30] M. Eizadjou, A.K. Talachi, H.D. Manesh, H.S. Shahabi, K. Janghorban, Compos. Sci. Technol. 68 (2008) 2003–2009.

[31] M. Alizadeh, M. Talebian, Mater. Sci. Eng. A 558 (2012) 331–337.

[32] C.Y. Liu, B. Zhang, P.F. Yu, R. Jing, M.Z. Ma, R.P. Liu, Mater. Sci. Eng. A 580 (2013) 36–40.

[33] D. Yadav, R. Bauri, Mater. Sci. Technol. 31 (2015) 494–500.

[34] S. Abraham, B.C. Pai, K.G. Satyanarayana, V.K. Vaidyan, J. Mater. Sci. 27 (1992) 3479–3486.

[35] B.B. Singh, M. Balasubramanian, J. Mater. Process. Technol. 209 (2009) 2104–2110.

[36] H. Okamoto, in: T.B. Massalski (Ed.), Binary Alloy Phase Diagrams, second ed., American Society for Metals, Metals Park, OH, 1990, pp. 234–235.

[37] R.S. Mishra, Z.Y. Ma, I. Charit, Mater. Sci. Eng. A 341 (2003) 307–310.

[38] D. Yadav, R. Bauri, Mater. Lett. 64 (2010) 664–667.

[39] A. Gholinia, P.B. Prangnell, M.V. Markushev, Acta Mater. 48 (2000) 1115–1130.

[40] A.P. Zhilyaev, B.K. Kim, J.A. Szpunar, M.D. Baró, T.G. Langdon, Mater. Sci. Eng. A 391 (2005) 377–389.

[41] R.Z. Valiev, Mater. Sci. Eng. A 234–236 (1997) 59–66.

[42] Y.G. Ko, D.H. Shin, K.T. Park, C.S. Lee, Scripta Mater. 54 (2006) 1785–1789.

[43] K.V. Jata, K.L. Semiatin, Scripta Mater. 43 (2000) 743–749.

[44] J.Q. Su, T.W. Nelson, R.S. Mishra, M.W. Mahoney, Acta Mater. 51 (2003) 713–729.

[45] T.R. McNelley, S. Swaminathan, J.Q. Su, Scripta Mater. 58 (2008) 349–354.

[46] J.Q. Su, T.W. Nelson, C.J. Sterling, Mater. Sci. Eng. A 405 (2005) 277–286.

[47] J.Q. Su, T.W. Nelson, C.J. Sterling, Philos. Mag. 86 (2006) 1–24.

[48] R.W. Fonda, J.F. Bingert, K.J. Colligan, Scripta Mater. 51 (2004) 243–248.

[49] L.S. Toth, J.J. Jonas, Scripta Metall. Mater. 27 (1992) 359–363.

[50] F.J. Humphreys, M. Hatherly, Recrystallization and Related Annealing Phenomena, second ed., Elsevier, Oxford, 2004.

[51] I. Charit, R.S. Mishra, Scripta Mater. 58 (2008) 367–371.

[52] T. Omori, T. Kusama, S. Kawata, I. Ohnuma, Y. Sutou, Y. Araki, et al., Science 341 (2013) 1500–1502.

[53] E.M. Taleff, N.A. Pedrazas, Science 341 (2013) 1461–1462.

[54] D. Yadav, R. Bauri, A. Kauffmann, J. Freudenberger, Metall. Mater. Trans. A 47 (2016) 4226–4238.

[55] Y. Morisada, H. Fujii, T. Nagaoka, M. Fukusumi, Mater. Sci. Eng. A 433 (2006) 50–54.

[56] A.D. Smigelskas, E.O. Kirkendall, Tran. Am. Inst. Min. Metall. Petro. Eng. 171 (1947) 130–142.

CHAPTER 5

Surface Composites by FSP

ABSTRACT

Surface engineering is becoming increasingly important for engineering components which involve surface interactions. The surface of such components is either protected, for example by protective coatings, or modified in a manner that suits the interactions at the surface. Surface hardening of components that are subjected to friction and consequent wear is a common industrial practice. While there are many surface hardening techniques in use they have their own limitations. Surface composites have been emerging as an attractive way to enhance the surface hardness and protect it against wear and tear. Friction stir processing (FSP) has emerged as an effective technique for surface modification and hardening. The technique also allows incorporation of hard ceramic reinforcement into the modified surface to further enhance the hardness. This chapter will deal with such surface composites made by FSP. It should be however, remembered that unless otherwise mentioned specifically, the distinction between surface and bulk composites made by FSP is not absolute and depends on how one perceives the depth to which particles are distributed and the intended application. The processing approaches for making bulk and surface composite by FSP, therefore, also remain the same. These approaches are detailed in Chapter 3, Processing Metal Matrix Composite (MMC) by FSP, and hence, the processing methods of surface composites will be briefly described here.

5.1 PROCESSING SURFACE COMPOSITES BY FSP

Fabrication of Al–SiC surface composite by FSP was first reported by Mishra et al. [1]. In their approach, SiC powder was added in methanol and then applied as a layer on the 5083 Al plate surface. FSP was then carried out over the dried layer to incorporate the SiC particles. A similar approach of pasting the reinforcement layer prior to FSP was adopted by some other researchers also for making surface composites [2,3]. This method however, has limitation in terms of

Metal Matrix Composites by Friction Stir Processing. DOI: http://dx.doi.org/10.1016/B978-0-12-813729-1.00005-X

confinement and depth of penetration of particles. Therefore other methods such as groove filling and drill-hole methods came into being for fabrication of surface composites. A large number of surface composites have been fabricated by the groove filling method due to its simplicity and better confinement of the particles in the groove [4–13]. Drilled holes on the plate have also been used as reservoir for the reinforcement powder [14–15].

Several other methods have also been used to fabricate a variety of surface composites by FSP. Al/CNT composite, for example, was fabricated by a combination of powder metallurgy (PM) consolidation and FSP [16]. The Al and carbon nanotube (CNT) powders were mixed in rotary mixer followed by cold compaction and hot pressing. The hot pressed billet was then subjected to FSP to fabricate the final composite. Liu et al. also used PM approach to fabricate in situ nanocomposites from Al–Mg–CuO powder mixture [17]. The constituent powders reacted during FSP to form nanometric MgO and Al_2Cu particles. Coating the surface of the plate with a composite or reacting species followed by FSP over the coating has also been utilized to make hard composite layers on the surface. Anvari et al. [18] used an air plasma spraying (APS) system to coat a 150 μm thick Cr_2O_3 layer on an Al 6061 plate which was then subjected to FSP. Cr_2O_3 was reduced to Al_2O_3 by a displacement reaction between Al and Cr_2O_3 due to the heat and stirring during FSP. The final composite contained a number of fine reinforcements (Cr_2O_3, $Al_{13}Cr_2$, and $Al_{11}Cr_2$) in the size range around 100 nm. The APS technique was also utilized to deposit a nanocomposite coating of Al–10% Al_2O_3, which was produced by mechanical alloying, on an AA2024 alloy plate [19]. The coated plate was subsequently subjected to FSP to fabricate nano surface composites. The average thickness of the surface composite layer (SCL) was about 600 μm with uniform distribution of Al_2O_3 particles. The microhardness of the composite layer (230 Hv) was significantly higher compared to that of the AA2024 substrate (90 Hv). Mazaheri et al. also used the coating-FSP combination to fabricate A356/Al_2O_3 surface composites [20]. In this study, the A356 chips and 5 vol.% Al_2O_3 powder particles were mixed and ball milled. The milled composite powder was deposited onto grit blasted A356-T6 plates by high velocity oxy-fuel (HVOF) spraying followed by FSP over the coating. FSP resulted in a well-bonded and defect free composite layer on the aluminum alloy substrate. Hodder et al. [21] used cold spraying to first coat Al$-$Al$_2$O$_3$ composite powder on AA6061 alloy and carried out FSP over it to make a SCL. The highest Al$_2$O$_3$ content in the composites was reported to be 48 wt.% after FSP when the coating material contained 90 wt.% Al$_2$O$_3$. The hardness of this coating increased from 85 Hv in as-coated condition to a maximum hardness of 137 Hv after FSP. A combination of coating and electrical discharge during FSP has also been used to fabricate surface composite [22]. Both width and depth of the reinforced zone increased with the application of the electrical current. The reinforced layer had a depth of 200 μm and a width of 5800 μm when processed without current, while the layer processed with electrical current was 1224 μm deep and 8000 μm wide.

5.2 EFFECT OF PROCESS VARIABLES AND TOOL DESIGN

The process variables for making composites by FSP primarily fall into three categories; (i) *machine variables*, (ii) *tool design*, and (iii) *material related variables* (material properties). The mechanical properties of the materials, viz. yield strength and hardness, influence the selection of process parameters. High-melting materials such as steel and Ti would need higher heat input and the rotation speed to traverse speed ratio (ω/v), which controls the heat input, is selected accordingly. The thermal conductivity of the material also plays a role as it governs the peak temperature experienced by the material during FSP. Materials with high thermal conductivity would dissipate the heat quickly and hence, more heat input is needed to obtain a defect free stir zone (SZ) [23,24].

The machine variables are essentially the rotation speed (ω) and traverse speed (v). The ω/v ratio determines the amount of heat input into the material during FSP [25]. There must be sufficient heat input to plasticize the material and provide proper stirring action for incorporation of the reinforcement. A higher rotational speed may be required in preparing surface composite by FSP for uniform distribution of the reinforcement. Material flow around the tool is enhanced with increasing rotational speed. It was shown by Kurt et al. that increasing rotational speed gives rise to better distribution of SiC particles in Al [26]. However, higher rotational speed may lead to grain growth due to higher heat input. It has been reported that an increase in rotational speed led to coarser grain size in AZ91/SiC surface composites [27].

In fabrication of AZ31/MWCNT surface composite, Morisada et al. varied the traverse speed from 25 to 100 mm/min while keeping the rotational speed constant at 1500 rpm [28]. At the highest traverse speed of 100 mm/min the CNTs were not dispersed uniformly. As the traverse speed was decreased to 50 mm/min the dispersion improved though there were still some regions with aggregated CNTs. The best distribution was obtained at 25 mm/min. It was attributed to higher heat input due to the lowering of traverse speed that provided a suitable viscosity to the AZ31 matrix. Shahraki et al. also reported that distribution of ZrO_2 particles in AA5083 alloy was not uniform at high-traverse and low-rotational speed due to insufficient heat input [29]. A combination of high rotational and low traverse speed gave rise to better particle distribution in compliance with the theory that such a combination (high ω/v ratio) gives higher heat input and hence, better stirring action in the SZ.

The temperature of the SZ and the exposure time are the important parameters in fabrication of in situ surface composite by FSP. Higher rotational speed will provide more heat input and lower traverse speed increases the high temperature exposure time for the in situ chemical reaction to occur efficiently. A higher rotation speed, however, can generate excessive heating. Chen et al., therefore, chose a lower rotation speed of 500 rpm and combined it with lower traverse speed to provide enough time for the chemical reaction to complete in the $Al-CeO_2$ system [30]. At lower traverse speed of 85 mm/min the reaction between Al and CeO_2 was completed producing particles of δ^*-Al_2O_3, a transition phase of alumina. On the other hand, the heat input at higher traverse speed (120 mm/min) was insufficient to complete the reaction and unreacted CeO_2 was left in the resultant in situ composite.

The distribution of particles plays a very important role on the properties of composites. An inhomogeneous distribution or particle clustering may lead to inferior properties. It has been reported that many a times a single pass of FSP may not be sufficient to homogeneously distribute the reinforcement [31]. The particle distribution generally improves with increasing number of passes. Asadi et al. showed that distribution of SiC and Al_2O_3 particles improved with number of FSP passes [32]. The Al_2O_3 particles formed clusters which were expanded in the material flow direction in the third pass though the cluster size reduced with the passes. The cluster size of the alumina particles decreased from 7 μm in the first pass to about 500 nm in the sixth pass. The distribution of SiC improved significantly by the fourth pass and at the eighth pass it was uniform with no strip like features which were observed at lower number of passes. Similarly, the cluster size of nano-Al_2O_3 is reported to reduce and the distribution improved with the number of FSP passes [11,33,34]. The grain size of the matrix is also reported to be decreasing with increasing number of passes [11,33,34]. This has been attributed to higher strain exerted on the material by increasing number of passes and formation of grains with high dislocation density that inhibited the grain growth [33]. The pinning effect of fine reinforcement particles also contributed to restrict the grain growth. It has also been suggested that as the particle distribution becomes more homogeneous with increasing passes, it creates more particle–matrix interfaces [35]. A higher grain boundary area is also generated due to reduction in the grain size during subsequent passes. As a result, the number of dislocation sources (interfaces) available during the later passes is much larger. Further deformation with increasing number of passes generates dislocations from these sources and thus enhances the extent of recrystallization giving rise to a finer grain size [35].

Tool geometry mainly relates to the shoulder diameter, shoulder features, and pin geometry that includes pin size, shape, and profile. The effect of tool geometry on material flow during FSP has been discussed in Section 3.2.2. Various types of pin geometries have been used in preparing surface composites by FSP [36]. Some of these pin profiles and pin sizes are shown schematically in Fig. 5.1. Faraji et al. prepared AZ91/Al_2O_3 surface composites by two types of pin geometries and reported that the grain size was finer and particle dispersion was better with the triangular tool compared to the square tool [37]. It was attributed to sharp corners of the triangular tool that resulted in vigorous stirring. Mahmoud et al. on the other hand, reported better distribution of SiC particles in Al/SiC composites with a square tool compared to cylindrical and triangular tools [38]. They also investigated the effect of probe size and profile (thread) on the particle distribution. A better particle distribution was reported in the composites processed by the threaded tool compared to the plain tool. A similar observation was made by Hashemi and Hussain for Al/TiN surface composites [13]. The threaded profile strongly influences the temperature distribution, materials flow velocity, and strain around the pin in the SZ [39].

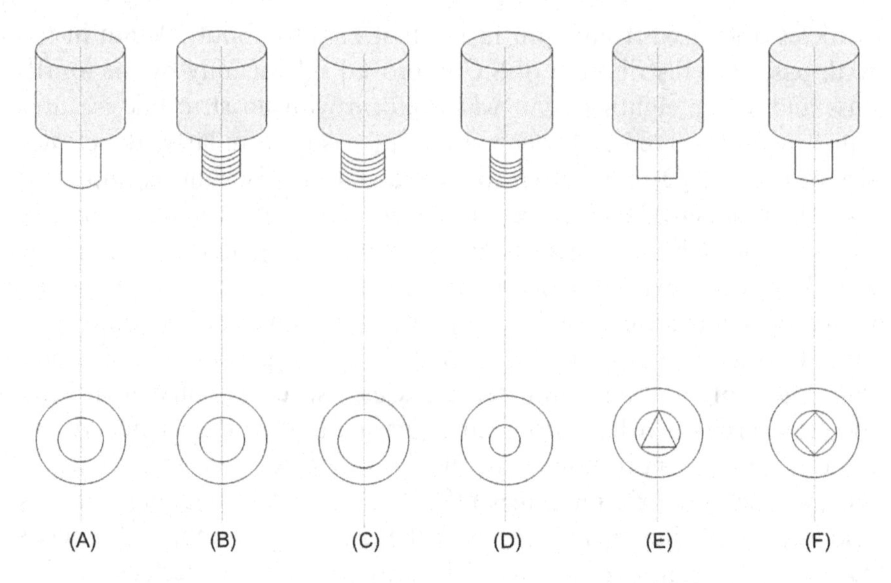

Figure 5.1 Different pin profiles used for making surface composites. (A) Circular, (B)–(D) circular with threads in different sizes, (E) triangular, and (F) square.

5.3 TYPES OF SURFACE COMPOSITES

Commonly used reinforcements in surface composites are Al_2O_3, SiC, TiC, B_4C, etc. These reinforcements are used in both micro- and nanometric size ranges. Nanostructures such as multiwalled carbon nanotubes (MWCNTs) have been also used as reinforcement. In situ formed fine intermetallic particles have also been used as reinforcements in many surface composites. Different types of surface composites are briefly reviewed in this section.

Surface nanocomposites form a major part of surface composites fabricated by FSP. Nanoparticles tend to agglomerate owing to their high surface area and therefore, dispersing them uniformly in the matrix is a challenge [40]. The synergistic effect of high temperature, severe plastic deformation, and materials mixing in FSP makes it an ideal processing tool to incorporate nanoparticles. A variety of nanoparticles have been incorporated by FSP in different matrices to form nanocomposites on the surface. A uniform dispersion of nanoparticles can be achieved by employing multiple FSP passes with right set of process parameters. Shafiei-Zarghani et al. processed nano-Al_2O_3 surface composite by multipass FSP [11,34]. The composite layer exhibited significant improvement in hardness. Moreover, the fine alumina particles also caused grain refinement in the matrix. Ghasemi-Kahrizsangi et al. processed Steel/TiC nano surface composites and achieved a uniform distribution of TiC particles in the steel matrix as shown in Fig. 5.2 [8].

CNTs exhibit exceptional properties and can be ideal reinforcement for surface composites with enhanced properties. The CNTs should be incorporated with minimum damage and good bonding with the matrix to derive the maximum advantage of the properties of CNTs. Johannes et al. investigated the survivability of single-walled CNTs during FSP [41]. Raman spectroscopy and SEM results indicated that the CNTs survived the thermal and stress cycles during FSP conducted with tool rotation speed of 400 rpm and traverse speed of 25.4 mm/min. Liu et al. studied the CNTs damage during multipass FSP of 2009Al–CNT billets consolidated by PM [42]. No obvious diameter change was observed in the CNTs as the number of FSP passes increased. The diameter of CNTs gradually changed from 11.8 nm to 10.6 nm between the first and the fifth FSP pass. This indicated that the CNTs did not suffer severe damage during FSP. The SEM micrographs of extracted CNTs after different FSP passes are shown in Fig. 5.3. The length of the CNTs, however, decreased with increasing number of FSP passes indicating cumulative damage to the CNTs by multiple passes of FSP. The processed composites (4.5 vol.% CNTs/2009Al) exhibited significant

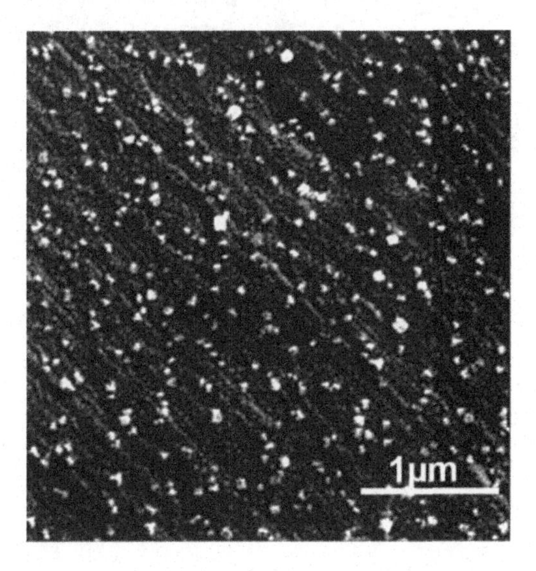

Figure 5.2 SEM micrograph showing distribution of nano-TiC particles in steel/TiC surface composite [8].

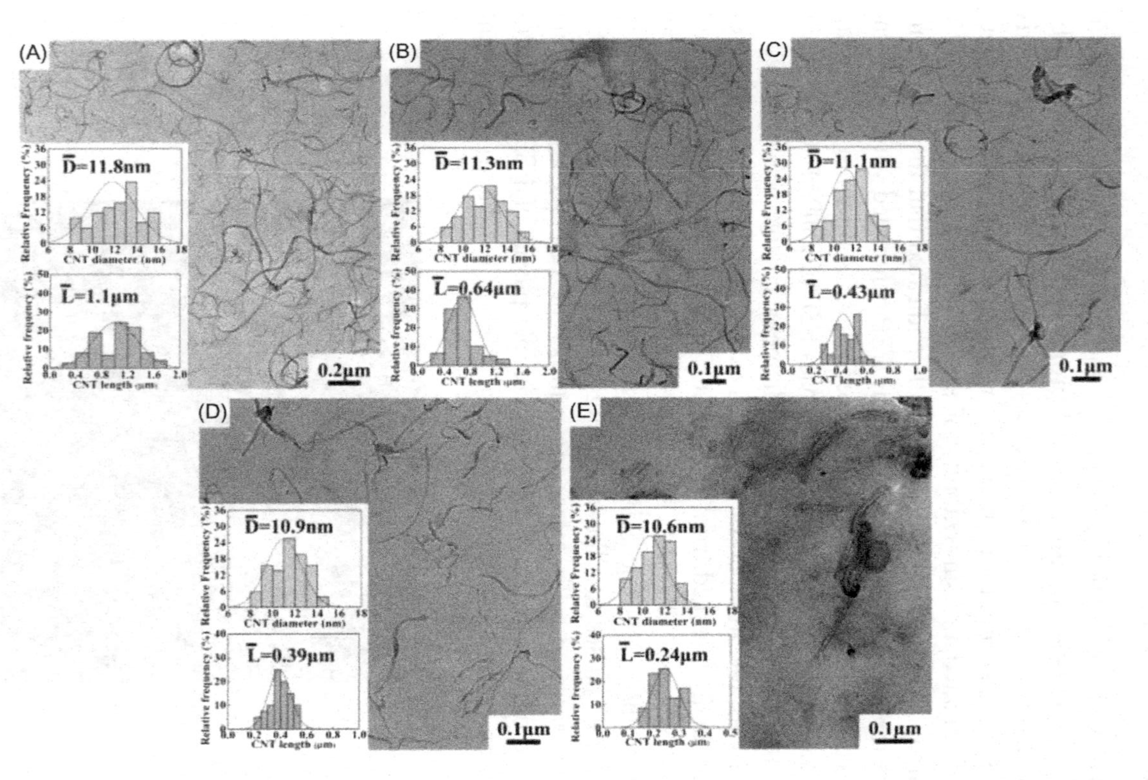

Figure 5.3 Morphologies and statistical lengths and diameters (inserts) of extracted CNTs at different numbers of FSP passes: (A) one pass, (B) two passes, (C) three passes, (D) four passes, and (E) five passes [42].

improvement in mechanical properties compared to the as-forged composites due to improved distribution of CNTs and densification during FSP. Zinati et al. incorporated MWCNTs in Polyamide 6 (PA 6) by FSP and developed a model based on Lagrangian incremental formulation to investigate the thermomechanical behavior [43]. The simulation results showed that the temperature was highest at the interface between tool shoulder and workpiece. The effective plastic strain was also highest on the top surface and the material shearing was higher in the advancing side compared to the retreating side.

Other nanoreinforcements such as nanohydroxyapatite have also been dispersed in metallic matrices by FSP. Farnoush et al. incorporated nanohydroxyapatite in Ti-6Al-4V by FSP [44]. The surface nanocomposite exhibited a hardness improvement of 33% at the top layer which was severely deformed during FSP. The depth of the processed layer reached to about 160 μm. Ratna Sunil et al. fabricated nanohydroxyapatite (nHA) reinforced AZ31 Mg based surface composite [45]. They observed that the corrosion resistance of the composite surface layer was better compared to as-received and FSPed AZ31 due to combined effect of grain refinement and nHA particles.

Polymer matrix nanocomposites have also been fabricated by FSP. Barmouz et al. reported fabrication of high density polyethylene (HDPE) based nanocomposite by incorporation of nanoclay particles through FSP [46]. The nanoclay particles were packed inside a groove on the HDPE sheet and subjected to FSP. The composite processed by FSP demonstrated a three-fold increment in hardness compared to similar composites processed by conventional melt mixing technique.

Surface in situ *composite* is another category of composite layer fabricated by FSP. In the in situ composites the reinforcement particles are generated in situ during the processing itself. In situ composites offer distinct advantages of thermodynamically stable reinforcements, better particle distribution, clean particle–matrix interface, and a good interfacial bonding. FSP has been proved to be an effective route of fabricating in situ composites as the frictional heat and the intense stirring action promote the chemical reaction that generates the reinforcement during the process. The in situ reaction is exothermic in nature and therefore, the heat released during the reaction makes the process self-sustaining. In Al based in situ composites, intermetallics are formed at the interface between Al and the reactive elements (Fe, Ni, Ti, etc.)

added to it. The exothermic heat released at the interface accelerates the reaction and it proceeds in a self-propagating manner producing fine intermetallic particles throughout the matrix. FSP not only distributes the particles homogeneously but can also refine the particle size [47]. The literature mainly reports fabrication of $Al-Al_3Ti$ [47–50], $Al-Al_3Ni$ [51], $Al-Al_3Fe$ [52], and $Al-Al_2Cu$ [53] composites. A major fraction of these composites has been produced from a combination of PM consolidation and FSP. Al and Ti powders, for example, were mixed and consolidated by powder compaction followed sintering or hot pressing [47,48]. The sintering temperature and time used might not have been sufficient to initiate or complete the reaction due to limited diffusivity. A higher temperature or longer holding time on the other hand, may lead to coarser intermetallic particles. The PM billet is subjected to FSP after sintering to initiate or enhance the reaction. FSP promotes the in situ reaction due to mechanical activation caused by severe plastic deformation during the process. The strain rate in the SZ is typically in the range of $10-100\ s^{-1}$ [54] and the strain can reach up to ~ 40 [55]. The severe deformation results in the breakup of the oxide film surrounding Al and thus presents atomically clean surfaces of Al to Ti in close contact, facilitating diffusion between them. A high density of dislocations is introduced due to the severe plastic deformation during FSP. Pipe diffusion can occur along the dislocation core. Pipe diffusion is three orders of magnitude higher than bulk diffusion at 400 °C in Al [56]. Hence, the diffusion between Al and Ti is expected to be substantially enhanced. As a result of all these factors, the Al–Ti reaction during FSP becomes interface-reaction controlled rather than diffusion controlled.

The other approach of making in situ composites is to pack the reacting powder in grooves made on the Al plate prior to FSP [49,51]. In this case also, the reaction may not be complete and multiple FSP passes are required to take the reaction to completion. For example, in an effort to make $Al-Al_3Ti-MgO$ composite TiO_2 particles were packed in grooves on Al–Mg plates and subsequently subjected to FSP [54]. More than 55% TiO_2 particles remained unreacted due to insufficient exposure time at the processing temperature during FSP and post-processing heat treatment was needed to complete the reaction [57]. Presence of Cu is shown to increase the reaction between Al and Ti to form more Al_3Ti. This has been attributed to formation of $Al-Al_2Cu$ eutectic which can accelerate the diffusion between Al and

Ti due to the presence of a liquid phase and thus enhancing the formation of the Al_3Ti phase [58].

Other systems such as Al–Cr–O and Al–CeO_2 have also been used to make surface in situ composites by FSP. FSP was applied on the plasma sprayed Cr_2O_3 coating on a 6061Al plate. Cr_2O_3 reacted in situ with Al during FSP and generated Al_2O_3 and other fine intermetallic reinforcements (Cr_2O_3, $Al_{13}Cr_2$, and $Al_{11}Cr_2$) [18]. In case of the Al–CeO_2 system, Al and CeO_2 reacted to form Al_2O_3 and $Al_{11}Ce_3$ [30]. Mg and Ti based in situ surface composites have been also fabricated by FSP. Abdollahi et al. fabricated AZ31 Mg based composites reinforced with Mg–Al–Ni based intermetallic particles. Ni particles were compressed in a groove made on the AZ31 plate having a nominal composition of 3.2% Al and 1% Zn. The in situ reactions during FSP generated complex intermetallics such as Mg_2Ni, Al_3Ni_2, and AlNi in the SZ volume [59]. Shamsipur et al. fabricated in situ Ti/TiN surface composites by subjecting a cp-Ti plate to FSP under nitrogen gas atmosphere [60]. The schematic of the set up for gas purging during FSP is shown in Fig. 5.4. It was reported that the TiN phase forms due to nitrogen diffusion and FSP distributed it in the SZ forming a Ti/TiN SCL. The transitional phases on the surface of gas-nitirided Ti sample can be as follows [61].

$$\alpha\text{-Ti} \rightarrow \alpha\text{-Ti(N)} \rightarrow Ti_2N \rightarrow TiN \tag{5.1}$$

An interstitial solid solution of N in α-Ti, i.e., α-Ti (N), can form at low nitrogen concentration. In conventional gas nitriding process it takes several hours for the TiN layer to form. The processing time in

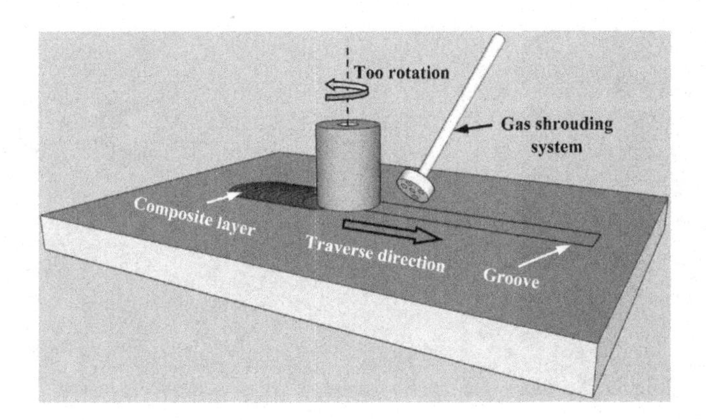

Figure 5.4 Schematic of FSP of Ti under flowing nitrogen gas.

FSP is much lower than that. However, during the nitrogen exposure in the FSP of Ti plate, the diffusion of nitrogen is enhanced by the large number of crystal defects generated due to the severe plastic deformation during FSP. The grain and subgrain boundaries also assist in diffusion. Moreover, new surface of the hot plasticized material is continuously exposed to nitrogen due to material flow and stirring action. As a result the nitrogen concentration in the metal increases leading to formation of the TiN phase which is dispersed as particles by the stirring action of FSP. Fig. 5.5 shows the microstructure of the Ti/TiN surface composite formed after four FSP passes. The hardness of this composite layer increased to ~972 HV which was about 6.4 times higher than that of the as-received cp-Ti substrate.

Surface hybrid composites contain more than one type of reinforcement. Use of more than one reinforcement comes from the need of satisfying certain property requirements. Hybrid composites combine the advantages of all the constituent reinforcements and hence, exhibit better properties compared to singly-reinforced composites [62]. The properties of the hybrid composites depend on the ratio of the volume fraction of reinforcement particles. For example, in the Ti/(TiB + TiC) hybrid composites a TiB:TiC ratio of 1:1 yielded higher strength and ductility compared to a ratio of 4:1 [62]. Hybrid composites have been processed by several methods such as PM [63], pressure infiltration [64], plasma spraying [65], and squeeze casting [66]. FSP is a surface modification process and can be used to incorporate a variety of reinforcing particles owing to the material mixing that happens during the process. Therefore FSP can be effectively used to fabricate surface

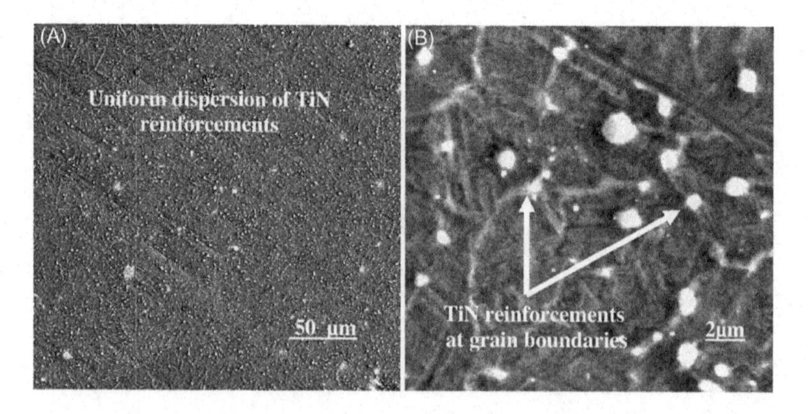

Figure 5.5 Microstructure of Ti/TiN layer produced after four FSP passes under flowing nitrogen gas [60].

hybrid composites. Mostafapour et al. studied the properties of AA5083/(Gr + Al_2O_3) hybrid composites fabricated by FSP [67]. The properties were evaluated as a function of the hybrid ratio (Gr:Al_2O_3). In this case also a 1:1 ratio (50% Gr:50% Al_2O_3) resulted in best strength. Increasing the graphite hybrid ratio (Gr.H.R) or decreasing the Al_2O_3 hybrid ratio (Al.H.R) led to lower strength and ductility. This was attributed to weaker graphite particles acting as crack initiation sites. As the Al_2O_3 content was increased or the Al.H.R was increased the interparticles distance between nano-alumina particles decreased and as a result the strength increased. Increasing the Al.H.R beyond 1:1 caused clustering of particles and as a result the interparticle spacing increased and the strength decreased. The wear rate, however, was lowest at a ratio of 3:1 (75% Gr) though the hardness decreased beyond 1:1 ratio. The lower wear rate in spite of decreasing hardness with increasing graphite content beyond 1:1 ratio was attributed to the lubricating property of graphite. Devaraju et al. also reported that increasing the graphite content up to a certain limit in AA6061/(Gr + SiC) hybrid composites decreases the hardness and wear rate [68]. Graphite acted as a solid lubricant that prevented direct metal to metal contact with the mating surface. Further increasing the graphite content resulted in lower wear resistance due to reduction in the fracture toughness of the composite that led to easier fracture during the wear process. Similarly, it has been shown that presence of MoS_2, which is a solid lubricant, can reduce the wear rate of surface composites. In the tribological study of A356/(MoS_2 + SiC) surface hybrid composites, it was found that the wear resistance of the hybrid composite was significantly higher compared to singly-reinforced A356/SiC composite [69]. It was attributed to the presence of MoS_2 particles in the mechanically mixed layer (MML) that led to a lubricating effect and lowered the wear. Mahmoud et al. fabricated surface Al/(SiC + Al_2O_3) hybrid composites with different weight ratios of SiC and Al_2O_3 [12]. In their case, an 80:20 ratio of SiC and Al_2O_3 resulted in higher hardness and lower wear resistance at 5 N load.

Various other reinforcement combinations have been also used in hybrid surface composites made by FSP. A combination of TiB_2 and Al_2O_3, for example, has been used as hybrid reinforcement in AA8026/TiB_2-Al_2O_3 surface composites [5]. Maxwell Rejil et al. processed AA6360/(TiC + B_4C) hybrid surface composites by two-pass FSP [9] and found the best wear properties at 1:1 ratio of TiC and B_4C.

Lu et al. explored a combination of nano-Al_2O_3 and CNT to fabricate surface composite on AZ31 Mg [70]. Kim et al. also used the same reinforcement combination to fabricate A356/(MWCNT + Al_2O_3) hybrid composite in an effort to overcome the difficulty of dispersing CNTs by combining it with another reinforcement which has better wettability with Al [71]. Hosseini et al. incorporated CNTs and nano-CeO_2 on 5083 Al surface and studied their effect on the mechanical properties and corrosion behavior [72]. While a combination of CNTs and CeO_2 in 3:1 proportion yielded maximum hardness and strength, a better resistance to pitting corrosion was obtained with only CeO_2 reinforcement.

5.4 WEAR BEHAVIOR OF SURFACE COMPOSITES

The operating environment of a large number of engineering components involves surface interactions and friction and wear of the material becomes an important concern in such applications. Wear of a material depends on various factors, such as material properties, the geometry of the contacting surfaces, environmental, and the operating conditions. However, the hardness of the material is historically considered to be the most important factor influencing the wear behavior of a material. Therefore increasing the hardness of the material is a widely accepted method in protection against wear and friction. The metal matrix composite (MMC) technology, wherein, the softer metallic matrix is reinforced with harder and stronger ceramic reinforcements, has become popular in that regard. A major drawback of MMCs, however, has been the low ductility which limited their widespread applications. Moreover, a hard surface layer often is the need rather than having a fully hardened bulk material. The ductility of the base material below the hard layer is also maintained in that case. Surface composites thus become very relevant in providing surface protection against wear and tear. FSP has emerged as an effective tool for making surface composites on a variety of substrate materials as described in the preceding sections. The wear behavior of such SCLs will be discussed in this section. However, before we proceed further into that it will be appropriate, at this juncture, to briefly review the wear mechanisms that are prevalent in MMCs.

In MMCs subjected to dry sliding against a harder counterface (e.g., hardened steel), there are mainly four types of wear: (i) *abrasive wear*, (ii) *adhesive wear*, (iii) *oxidative wear*, and (iv) *delamination*

[73–75]. The two main parameters influencing the occurrence or prevalence of a particular type of wear, are the applied load and the sliding distance [74]. At low loads, the couterface is quickly abraded by hard asperities until a steady state is reached. This is called running-in which typically occurs in the first 50 m of sliding. The variation in friction coefficient and volumetric wear of 6061Al/Al_2O_3 composites under a normal stress of 0.69 MPa is shown in Fig. 5.6 [74]. Regime "A" is the running-in period in which the friction coefficient increases steadily to a maximum and then drops off to a steady state. Consequently, the volumetric wear increases rapidly initially and then slows down to a steady state. The main reason behind the steep rise in the wear rate in the early period (regime A) is that the ceramic particles protrude out and abrade the steel (counterface) surface like a cutting tool. At the same time, the asperities from the steel surface abrade the sample surface and form grooves. The plowing of the surfaces due to this abrasion was modeled by Tabor [76] and was shown that the plowing component of friction, μ_p, depends on the depth of penetration (for a spherical asperity) or asperity sharpness (for conical asperities).

After the short period of abrasion the volumetric wear reaches a steady state (Regime "B" in Fig. 5.6). The examination of the worn surfaces and wear debris produced in this regime revealed formation of oxides. The temperature rises due to the frictional heat and oxidation of the steel surface readily occurs. The oxide particles or an oxide film lubricate the surface and bring down the friction coefficient and the wear

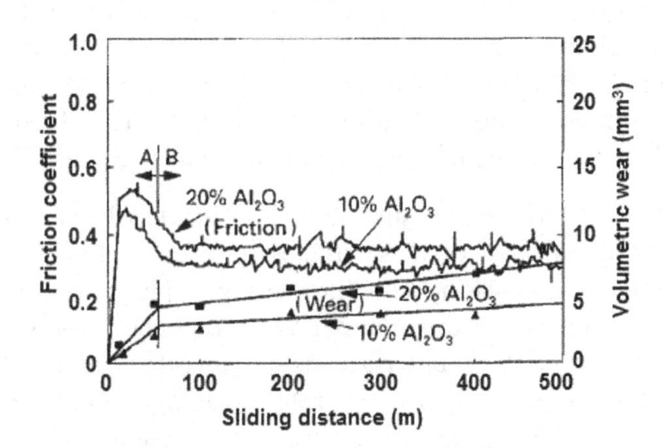

Figure 5.6 Variation of friction coefficient and volumetric wear with sliding distance for 6061Al/Al_2O_3 composites at a normal stress of 0.69 MPa [74].

loss [74]. With increasing sliding distance and or load, the asperities deform and the mating surfaces make adhesive contact. The deformed asperities generate wear particles which are typically flaky type [74,75]. Cross-transfer of material from the sample to the steel surface and vice versa also takes place due to the adhesion. Therefore adhesive wear can be characterized by flaky type wear debris and presence of the counterface material on the worn surface of the sample. The friction coefficient and volumetric wear increases due to material removal in this regime. The generation of wear particles transforms the wear from two-body to a three-body abrasion wear. It may be also noted here that a tribolayer forms just beneath the worn surface due to severe deformation and material flow and mixing. This layer is also known as mechanically mixed layer (MML) and consists of materials from both the mating surfaces [77]. The MML layer also contains finer fragmented reinforcement particles in case of MMCs. At high loads subsurface cracks grow due to large strain. In MMCs, the crack can initiate at the particle–matrix interfaces. As the worn material is removed from the surface layer the crack approaches the surface and causes material removal by delamination. Delamination wear causes excessive material removal in chunks leading to entry in the severe wear regime.

Al low loads, MMCs and their unreinforced counter parts both show mild wear and their wear rates do not differ significantly. At higher loads, on the other hand, the wear conditions change from the usual two-body abrasion to more severe delamination wear. In the severe wear regime, composites generally exhibit better wear performance compared to the unreinforced alloy. The transition from mild to severe wear depends on the load. The transition to severe wear is not suppressed by the load-bearing reinforcement in the composites rather it is delayed to higher loads [73,75]. The transition load is generally found to increase with increasing reinforcement content [73].

Wear behavior of several surface composites has been investigated [2,5–13,51,70]. In most of these studies the wear behavior has been evaluated under dry sliding condition using a pin-on-disc apparatus wherein the sample is used as the pin and a hardened steel disc (HRC ~60–65) is used as the counterface. The wear rate of SCL is found to be lower than the unreinforced substrate satisfying the requirement for which they are formed. The better wear performance of the SCL is generally attributed to the dispersion of ceramic particles and concomitant increment in the hardness. The relationship between hardness and wear rate is given by the Archard's equation [78].

$$V = \kappa \frac{NS}{H} \tag{5.2}$$

where V is the volume loss due to wear, N is the normal load, S is the sliding distance, H is the hardness, and κ is a dimensionless constant known as wear coefficient. The wear loss is inversely proportional to the hardness. Therefore as the hardness increases due to particle addition, the wear rate decreases. The grain refinement due to FSP also contributes to the hardening. The load-bearing ability of the stronger ceramic particles also play a role in lowering the wear of the composite [79]. The particles bear the load during the wear process and reduce the direct contact between the sample and the disc (counterface).

The wear behavior of composites, however, can be far more complicated than the simple empirical relation given in Eq. (5.2) due to various other factors such as particle distribution, particle size, defects, etc. [75]. For example, Devaraju et al. [6] found that wear rate of 6061Al/SiC–Al_2O_3 surface composite was lower at 8 vol.% SiC and 2 vol.% Al_2O_3 which were uniformly distributed. Significant reduction in the wear rate of the SCL has also been reported for AA8026 based surface composites reinforced with uniformly dispersed TiB_2 and Al_2O_3 particles [5]. However, as the volume fraction of the reinforcements increased the wear performance of the 6061Al/SiC–Al_2O_3 composite layer degraded. This was attributed to particle pull out possibly related to clustering effect. Various wear mechanisms also play a role in deciding the wear behavior of composites. In the nano-Al_2O_3 and CNT reinforced AZ31 surface composite the wear mechanism was reported to be abrasive wear at low loads (1.3 MPa) [70]. The worn surface showed many parallel grooves and material pile up at the edge of the grooves. The energy-dispersive X-ray spectroscopy (EDS) analysis revealed the presence of C, O, Mg, and Al, all of which came from the composite layer indicating no material transfer from the steel disc. All these aspects indicate towards occurrence of abrasive type wear. At higher load (3.25 MPa), on the other hand, there were cracks perpendicular or parallel to the wear direction and the EDS analysis of the subsurface layer showed presence of Fe and Cr in addition to the constituents of the composite material (Fig. 5.7). This indicates the formation of a MML. The cracks in the subsurface led to detachment of the MML giving rise to

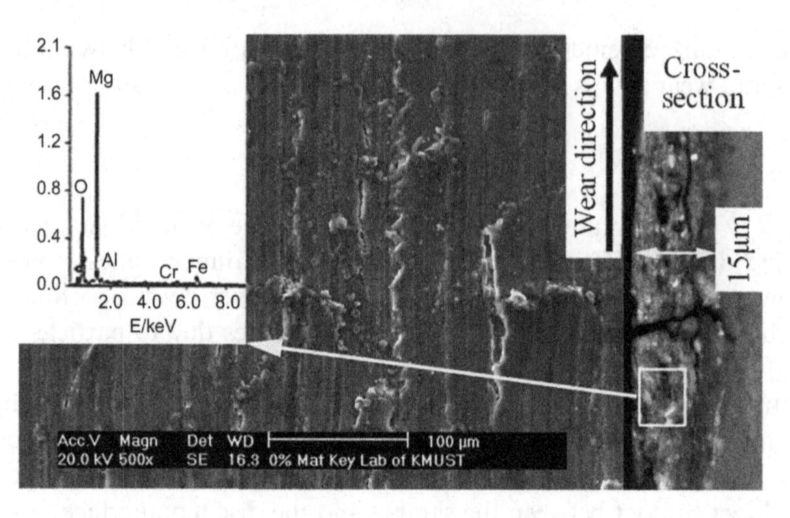

Figure 5.7 SEM images of worn surface and subsurface of AZ31/Al_2O_3–CNT SCL at 3.25 MPa [70].

delamination type of wear. Hashemi and Hussain reported that wear on the Al/TiN surface composite was controlled by abrasion and adhesion while the dominant wear mechanism in the base metal was adhesive wear [13]. Formation of a tribo film between the mating surfaces during wear has also been shown to be a controlling factor for wear of composite layers. In the wear test of Al/SiC–Al_2O_3 surface hybrid composites, for example, Mahmoud et al. report that for the composite having 20 % relative ratio of Al_2O_3 (80% SiC + 20% Al_2O_3) the pulled-out Al_2O_3 particles formed a thin tribo film over the worn surface providing a lubricating effect and thus, reducing the friction coefficient and the wear rate of the composite at 5 N load [12].

The applied load also has a significant influence of the wear behavior of composites. It has been reported that there exists a critical load beyond which the wear changes from mild wear to severe wear [73,75]. Shafiei-Zarghani et al. evaluated the wear performance of Al/Al_2O_3 nanocomposite surface layer at three different loads (20, 40, and 60 N) [11]. The steady state wear rate of the unreinforced base metal increased with increasing load (Fig. 5.8). The base metal exhibited a two-stage wear. The first stage wear up to a sliding distance of about 200 m was characterized by low wear rate followed by a high wear regime where the wear rate was higher by one to two orders of magnitude compared to the first stage. The two-stage wear trend was evident in the friction coefficient plotted as a function of sliding distance. The friction coefficient increased rapidly from 0.2 to 0.35 in the beginning

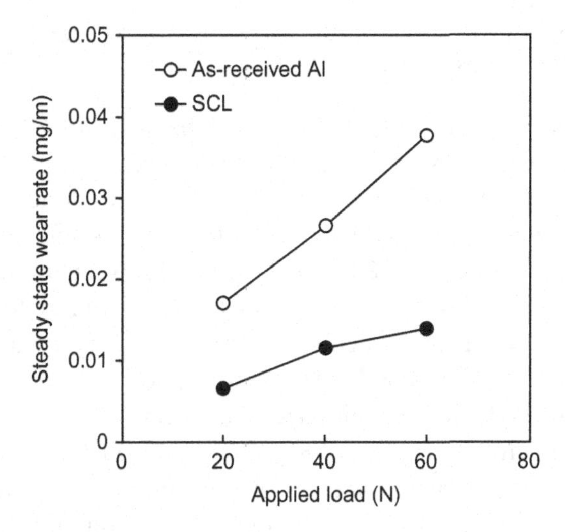

Figure 5.8 Steady state wear rate as a function of applied load for the as-received Al substrate and Al/Al_2O_3 nanocomposite SCL fabricated using four FSP passes [11].

within first 200 m of sliding followed by a jump to about 0.5 to 0.7 depending on the load and then settling down to a steady state. The sliding counterface could penetrate and cut the softer metal creating grooves and a number of pits (damaged spots) on the surface. The pits were found to be deeper and larger as the load was increased from 20 to 40 N. In addition, substantial wear debris were observed in and around the pits. The introduction of such wear particles converted the wear mode from two-body to three-body abrasion which increased the wear rate. At further higher load of 60 N irregular plastic flow lines were found on the worn surface indicating extensive plastic deformation during the wear process. Extensive pitting and local delamination was also observed at 60 N. This indicates that the wear mechanism changed from mild to severe wear with increasing load. The SCL, on the other hand, did not display the two-stage wear. The grooves formed on the SCL were narrower and pits were smaller compared to those on the base Al surface indicating mild wear. The wear rate of the SCL increased with increasing load just like the unreinforced metal. However, the wear rates of the SCL were substantially lower than the base metal and the difference in the wear rates was more significant at higher load (Fig. 5.8), an observation reported by others as well [80]. The incorporation of Al_2O_3 particles strengthened the matrix and provided protection against wear. A combination of abrasive (initial stage) and adhesive wear (later stage) was the dominant wear mechanism in all applied loads in the base metal whereas in case of the SCL, the wear mode was abrasive at 20 and 40 N loads and changed to predominantly adhesive at 60 N.

Shyam Kumar et al. evaluated the wear behavior of 5083 Al/W SCL as a function of load [81]. The unreinforced and friction stir processed (FSPed) 5083 Al alloy displayed mild wear at low load (25 N) but a transition to severe wear happened at higher loads (50 and 75 N). In contrast, the 5083 Al/W SCL displayed mild wear in all the three applied loads. The difference in the wear rate between the composite and the unreinforced alloy increased with increasing load in this case also. A significantly improved wear performance even compared to the fine-grained FSPed alloy highlighted the effectiveness of the tungsten particle reinforcement in providing protection against wear. This was also evident from the worn surfaces of the three samples at 50 N load. As observed in Fig. 5.9A and B, respectively, the unreinforced- and the FSPed-alloys show the formation of deep craters on the worn surface

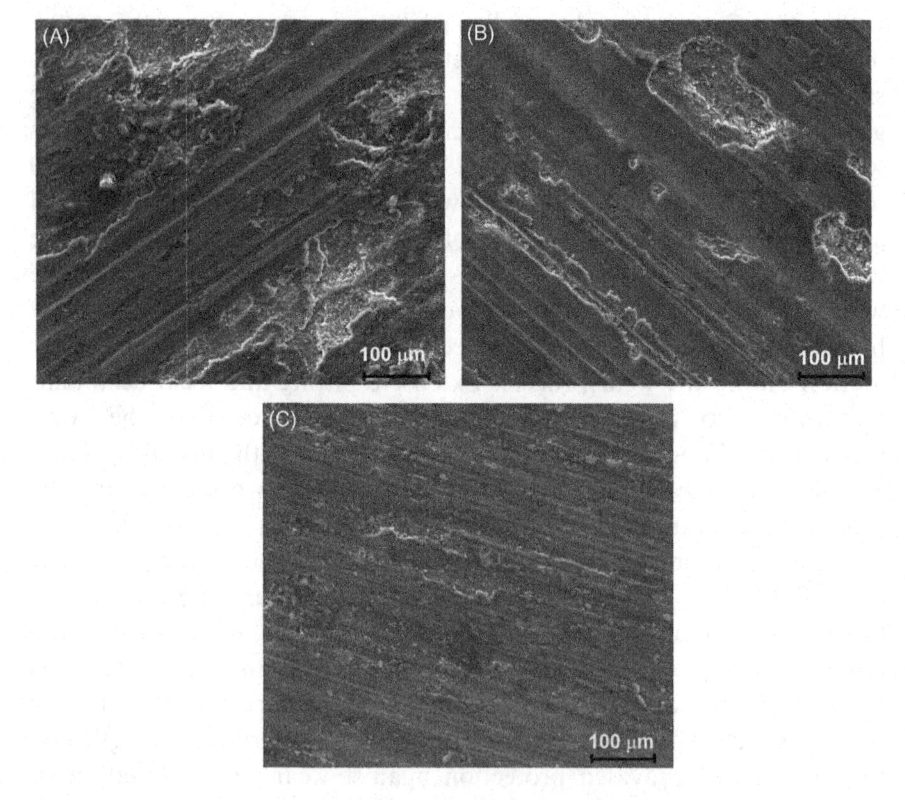

Figure 5.9 Worn surfaces of (A) base 5083 Al, (B) FSPed 5083 Al, and (C) 5083 Al/W SCL at 50 N load.

indicating material removal by delamination. The worn surface of the 5083 Al/W composite, on the other hand, does not show any extensive damage indicating occurrence of mild wear.

In some studies solid lubricants like graphite and MoS_2 have been used as a reinforcement/hybrid reinforcement to improve the wear resistance of SCL [2,4,7,67]. As the lubricant particles are smeared over the surface during the wear process, the friction between the mating surfaces reduce owing to the lubricating property of the particles. The particles also reduce the metal–metal contact points during the sliding process [4].

REFERENCES

[1] R.S. Mishra, Z.Y. Ma, I. Charit, Mater. Sci. Eng. A 341 (2002) 307–310.

[2] M. Raaft, T.S. Mahmoud, H.M. Zakaria, T.A. Khalifa, Mater. Sci. Eng. A 528 (2011) 5741–5746.

[3] R.M. Miranda, T.G. Santos, J. Gandra, N. Lopes, R.J.C. Silva, J. Mater. Process. Technol. 213 (2013) 1609–1615.

[4] H. Sarmadi, A.H. Kokabi, S.M. Seyed Reihani, Wear 304 (2013) 1–12.

[5] H. Eskandari, R. Taheri, F. Khodabakhshi, Mater. Sci. Eng. A 660 (2016) 84–96.

[6] A. Devaraju, A. Kumar, A. Kumaraswamy, B. Kotiveerachari, Mater. Des. 51 (2013) 331–341.

[7] S. Soleymani, A. Abdollah-zadeh, S.A. Alidokht, Wear 278–279 (2012) 41–47.

[8] A. Ghasemi-Kahrizsangi, S.F. Kashani-Bozorg, Surf. Coat. Technol. 209 (2012) 15–22.

[9] C. Maxwell Rejil, I. Dinaharan, S.J. Vijay, N. Murugan, Mater. Sci. Eng. A 552 (2012) 336–344.

[10] J. Qu, H. Xu, Z. Feng, D.A. Frederick, L. An, H. Heinrich, Wear 271 (2011) 1940–1945.

[11] A. Shafiei-Zarghani, S.F. Kashani-Bozorg, A. Zarei-Hanzaki, Wear 270 (2011) 403–412.

[12] E.R.I. Mahmoud, M. Takahashi, T. Shibayanagi, K. Ikeuchi, Wear 268 (2010) 1111–1121.

[13] R. Hashemi, G. Hussain, Wear 324–325 (2015) 45–54.

[14] M. Yang, C. Xu, C. Wu, K.-C. Lin, Y.J. Chao, L. An, J. Mater. Sci. 45 (2010) 4431–4438.

[15] B. Li, Y. Shen, L. Luo, W. Hu, Mater. Sci. Eng. A 574 (2013) 75–85.

[16] G.L. You, N.J. Ho, P.W. Kao, Mater. Lett. 100 (2013) 219–222.

[17] Z.Y. Liu, B.L. Xiao, W.G. Wang, Z.Y. Ma, J. Mater. Sci. Technol. 30 (2014) 649–655.

[18] S.R. Anvari, F. Karimzadeh, M.H. Enayati, J. Alloys Compd. 562 (2013) 48–55.

[19] B. Zahmatkesh, M.H. Enayati, Mater. Sci. Eng. A 527 (2010) 6734–6740.

[20] Y. Mazaheri, F. Karimzadeh, M.H. Enayati, J. Mater. Process. Technol. 211 (2011) 1614–1619.

[21] K.J. Hodder, H. Izadi, A.G. McDonald, A.P. Gerlich, Mater. Sci. Eng. A 556 (2012) 114–121.

[22] T.G. Santos, N. Lopes, M. Machado, P. Vilac, R.M. Miranda, J. Mater. Process. Technol. 216 (2015) 375–380.

[23] M.Z.H. Khandkar, J.A. Khan, A.P. Reynolds, Sci. Technol. Weld. Join. 8 (2003) 165–174.

[24] N. Xu, R. Ueji, H. Fujii, Mater. Sci. Eng. A 610 (2014) 132–138.

[25] R.S. Mishra, Z.Y. Ma, Mater. Sci. Eng. R 50 (2005) 1–78.

[26] A. Kurt, I. Uygur, E. Cete, J. Mater. Process. Technol. 211 (2011) 313–317.

[27] P. Asadi, G. Faraji, M.K. Besharati, Int. J. Adv. Manuf. Technol. 51 (2010) 247–260.

[28] Y. Morisada, H. Fujii, T. Nagaoka, M. Fukusumi, Mater. Sci. Eng. A 419 (2006) 344–348.

[29] S. Shahraki, S. Khorasani, R.A. Behnagh, Y. Fotouhi, H. Bisadi, Metall. Mater. Trans. B 44 (2013) 1546–1553.

[30] P. Chen, P.W. Kao, L.W. Chang, N.J. Ho, Metall. Mater. Trans. A 41 (2010) 513–522.

[31] D.K. Lim, T. Shibayanagi, A.P. Gerlich, Mater. Sci. Eng. A 507 (2009) 194–199.

[32] P. Asadi, G. Faraji, A. Masoumi, M.K.B. Givi, Metall. Mater. Trans. A 42 (2011) 2820–2832.

[33] M. Sharifitabar, A. Sarani, S. Khorshahian, M. Shafiee Afarani, Mater. Des. 32 (2011) 4164–4172.

[34] A. Shafiei-Zarghani, S.F. Kashani-Bozorg, A. Zarei-Hanzaki, Mater. Sci. Eng. A 500 (2009) 84–91.

[35] R. Bauri, D. Yadav, G. Suhas, Mater. Sci. Eng. A 528 (2011) 4732–4739.

[36] V. Sharma, U. Prakash, B.V. Manoj Kumar, J. Mater. Process. Technol. 224 (2015) 117–134.

[37] G. Faraji, O. Dastani, S.A.A.A. Mousavi, J. Mater. Eng. Perform. 20 (2011) 1583–1590.

[38] E.R.I. Mahmoud, M. Takahashi, T. Shibayanagi, K. Ikeuchi, Sci. Technol. Weld. Join. 14 (2009) 413–425.

[39] Z. Yu, W. Zhang, H. Choo, Z. Feng, Metall. Mater. Trans. A 43 (2011) 724–737.

[40] S.C. Tjong, Adv. Eng. Mater. 9 (2007) 639–652.

[41] L.B. Johannes, L.L. Yowell, E. Sosa, S. Arepalli, R.S. Mishra, Nanotechnology 17 (2006) 3081–3084.

[42] Z.Y. Liu, B.L. Xiao, W.G. Wang, Z.Y. Ma, Carbon 69 (2014) 264–274.

[43] R.F. Zinati, M.R. Razfar, H. Nazockdast, J. Mater. Process. Technol. 214 (2014) 2300–2315.

[44] H. Farnoush, A. Sadeghi, A. Abdi Bastami, F. Moztarzadeh, J. Aghazadeh Mohandesi, Ceram. Int. 39 (2013) 1477–1483.

[45] B. Ratna Sunil, T.S. Sampath Kumar, U. Chakkingal, V. Nandakumar, M. Doble, J. Mater. Sci. Mater. Med. 25 (2014) 975–988.

[46] M. Barmouz, J. Seyfi, M.K.B. Givi, I. Hejazi, S.M. Davachi, Mater. Sci. Eng. A 528 (2011) 3003–3006.

[47] Q. Zhang, B.L. Xiao, D. Wang, Z.Y. Ma, Mater. Chem. Phys. 130 (2011) 1109–1117.

[48] C.J. Hsu, P.W. Kao, N.J. Ho, Mater. Lett. 61 (2007) 1315–1318.

[49] F. Khodabakhshi, A. Simchi, A.H. Kokabi, A.P. Gerlich, Mater. Charact. 108 (2015) 102–144.

[50] A. Rezaei, H.R.M. Hosseini, Mater. Sci. Eng. A 689 (2017) 166–175.

[51] M. Golmohammadi, M. Atapour, A. Ashrafi, Mater. Des. 85 (2015) 471–482.

[52] I.S. Lee, P.W. Kao, N.J. Ho, Intermetallics 16 (2008) 1104–1108.

[53] E.R.I. Mahmoud, A.M.A. Al-qozaim, Arab. J. Sci. Eng. 41 (2016) 1757–1769.

[54] C.I. Chang, C.J. Lee, J.C. Huang, Scripta Mater 51 (2004) 509–514.

[55] P. Heurtier, C. Desrayaud, F. Montheillet, Mater. Sci. Forum 396–402 (2002) 1537–1542.

[56] M. Legros, G. Dehm, E. Arzt, T.J. Balk, Science 319 (2008) 1646–1649.

[57] F. Khodabakhshi, A. Simchi, A.H. Kokabi, A.P. Gerlich, M. Nosko, Mater. Des. 63 (2014) 30–41.

[58] Q. Zhang, B.L. Xiao, Z.Y. Ma, Intermetallics 40 (2013) 36–44.

[59] S.H. Abdollahi, F. Karimzadeh, M.H. Enayati, J. Alloys Compd. 623 (2015) 335–341.

[60] A. Shamsipur, S.F. Kashani-Bozorg, A. Zarei-Hanzaki, Surf. Coat. Technol. 218 (2013) 62–70.

[61] A. Zhecheva, W. Sha, S. Malinov, A. Long, Surf. Coat. Technol. 200 (2005) 2192–2207.

[62] W.J. Lu, D. Zhang, X.N. Zhang, R.J. Wu, T. Sakata, H. Mori, Mater. Sci. Eng. A 311 (2001) 142–150.

[63] S.C. Tjong, K.C. Lau, S.Q. Wu, Metall. Mater. Trans. A 30A (1999) 2551–2555.

[64] Z.M. Du, J.P. Li, J. Mater. Process. Technol. 151 (2004) 298–301.

[65] M. Gui, B. Kang, Metall. Mater. Trans. A 32A (2001) 2383–2392.

[66] H. Fu, K. Han, J. Song, Wear 256 (2004) 705–713.

[67] A. Mostafapour Asl, S.T. Khandani, Mater. Sci. Eng. A 559 (2013) 549–557.

[68] A. Devaraju, A. Kumar, B. Kotiveerachari, Mater. Des. 45 (2013) 576–585.

[69] S.A. Alidokht, A. Abdollah-zadeh, S. Soleymani, H. Assadi, Mater. Des. 32 (2011) 2727–2733.

[70] D. Lu, Y. Jiang, R. Zhou, Wear 305 (2013) 286–290.

[71] H.H. Kim, J.S.S. Babu, C.G. Kang, Mater. Sci. Eng. A 573 (2013) 92–99.

[72] S.A. Hosseini, K. Ranjbar, R. Dehmolaei, A.R. Amirani, J. Alloys Compd. 622 (2015) 725–733.

[73] A.P. Sannino, H.J. Rack, Wear 189 (1995) 1–19.

[74] Z.F. Zhang, L.C. Zhang, Y.W. Mai, J. Mater. Sci. 30 (1995) 1961–1966.

[75] J. Singh, A. Chauhan, Ceram. Int. 42 (2016) 56–81.

[76] D. Tabor, Friction and wear: development over the last fifty years, in: Proc. Int. Conf. Tribology – 50 Years On, 1987, London. Pub. Porc. Inst. Mech. Engr., C245, pp. 157–172.

[77] B. Venkataraman, G. Sundararajan, Acta Mater. 44 (1996) 461–473.

[78] J.F. Archard, Wear theory and mechanisms, in: M.B. Peterson, W.O. Winer (Eds.), Wear Control Handbook, ASME, New York, 1980, pp. 5–80.

[79] R. Bauri, M.K. Surappa, Wear 265 (2008) 1756–1766.

[80] A. Ghasemi-Kahrizsangi, S.F. Kashani-Bozorg, M. Moshref-Javadi, Surf. Coat. Technol. 276 (2015) 507–515.

[81] C.N. Shyam Kumar, R. Bauri, D. Yadav, Tribo. Int. 101 (2016) 284–290.

CHAPTER 6

Summary and Future Direction

6.1 SUMMARY

Friction stir processing (FSP) has emerged as an effective tool for grain refinement and microstructure modification. It has now been also demonstrated that FSP is an attractive method to process composites. A wide range of reinforcements that include ceramic and metallic particles, carbon nanostructures have been incorporated in metallic matrices by FSP. This process has proved its potential in fabricating both bulk and surface composites. Nanocomposites and hybrid composites have been also fabricated by FSP. Different approaches such as groove filling, drill-hole, powder metallurgy and FSP combination, and in situ method have been followed to process a variety of composites. A major concern in fabricating the metallic matrix composites (MMCs) has been the inhomogeneity of particle distribution. A number of tool designs have emerged in order to distribute the particles uniformly. It has also been demonstrated how the process parameters affect the microstructure and properties of the processed composites. Both single-pass and multi-pass FSP have been explored and it has been found that the particle distribution generally improves with the number of passes. FSP has been also found useful in homogenizing the microstructure of as-cast and other MMCs that suffer from microstructural inhomogeneity such as particle clustering.

The processed composites are generally found to have better mechanical properties compared to the unreinforced base matrix. Grain refinement and Orowan strengthening are found to be the major strengthening mechanisms in these composites. Much of these properties depend on the particle distribution and how the particles affect the microstructure evolution during the process. A number of studies have shown the role of reinforcement particles on the microstructure evolution and a general trend of finer grain size due to the pinning effect of the particles has been found. Some of the studies have also tried to establish a microstructure–property correlation in the composites processed by FSP. The surface composites fabricated by FSP exhibited

Metal Matrix Composites by Friction Stir Processing. DOI: http://dx.doi.org/10.1016/B978-0-12-813729-1.00006-1

higher hardness and hence better wear resistance compared to the monolithic base material. Different forms of wear are shown to be operative depending on the applied load and sliding distance during the wear process of the surface composites. The load-bearing ability and the lubricating effect of the reinforcements have greatly reduced the wear rate of the surface layer composites.

6.2 FUTURE DIRECTION

Chapter 4, Processing Nonequilibrium Composite (NMMC) by FSP, of this book reports a new type of composite that was named as non-equilibrium composite being processed by FSP. Such studies highlight the potential of FSP toward the development of new materials. The process offers the possibility of incorporating new types of reinforcements. Some reports of fiber incorporation, for example, has been already found and the results are encouraging [1,2]. The possibility of fabricating functionally graded composites by FSP has been also reported [3]. This kind of graded microstructure can be created on the surface by variety of other reinforcements as well.

So far, FSP has been primarily applied to softer and lower-melting metals like Al and Mg. Though there are some reports on FSP of stronger and high-melting materials like Ti and steel, the studies are limited. Therefore, it will be interesting to extend the potential of FSP to process these stronger materials and their composites. The tool wear can be a concern for FSP of such materials and hence the tool material and tool design will take the central role in such cases. Tools made of polycrystalline cubic boron nitride (PCBN), tungsten carbide, or tungsten-based harder alloys are recommended. Then there are polymer-based materials that can be explored to develop nonmetallic composites by FSP. The preliminary reports on FSP of polymer and processing polymer-based composites by FSP are promising.

FSP is primarily a surface intensive process and hence can be potentially applied for location-specific property improvement on the surface of engineering components. In a recent study, Al surface was alloyed with Zn by FSP and this enhanced the hardness of the surface without affecting the bulk substrate below [4]. Other high-solubility metals can be also incorporated by FSP for surface alloying. In a nut shell, it can be said that FSP has the potential toward opening up a new frontier of surface engineering.

REFERENCES

[1] S.M. Arab, S. Karimi, S.A.J. Jahromi, S. Javadpour, S.M. Zebarjad, Mater. Sci. Eng. A 632 (2015) 50–57.

[2] A. Mertens, A. Simar, J. Adrien, E. Maire, H.-M. Montrieux, F. Delannay, et al., Mater. Charact. 107 (2015) 125–133.

[3] M. Salehi, H. Farnoush, J.A. Mohandesi, Mater. Des. 63 (2014) 419–426.

[4] D. Yadav, Ph.D. thesis, Indian Institute Technology Madras, India, 2015.

Printed in the United States
By Bookmasters